And the Sun Stood Still

BY THE SAME AUTHOR

A More Perfect Heaven

Longitude

The Illustrated Longitude
(with William J. H. Andrews)

Galileo's Daughter

The Planets

Letters to Father
(translated and annotated)

And the Sun Stood Still

A Play in Two Acts

DAVA SOBEL

BLOOMSBURY
NEW YORK · LONDON · OXFORD · NEW DELHI · SYDNEY

Bloomsbury USA
An imprint of Bloomsbury Publishing Plc

1385 Broadway	50 Bedford Square
New York	London
NY 10018	WC1B 3DP
USA	UK

www.bloomsbury.com

BLOOMSBURY and the Diana logo are trademarks of
Bloomsbury Publishing Plc

First published 2016

© Dava Sobel, 2016

Art credits: Page viii, Bayerische Staatsbibliothek, Munich, Cod. Lat. #27003, folio 33. Page xvi, maps © Jeffrey L. Ward, 2011. Page 90, Captain Dariusz Zajdel, M.A., Central Forensic Laboratory of the Polish Police/AFP Getty Images.

All rights reserved. No part of this publication may be reproduced or transmitted in any form or by any means, electronic or mechanical, including photocopying, recording, or any information storage or retrieval system, without prior permission in writing from the publishers.

No responsibility for loss caused to any individual or organization acting on or refraining from action as a result of the material in this publication can be accepted by Bloomsbury or the author.

ISBN: HB: 978-0-80271-694-1
ePub: 978-0-80277-802-4

Library of Congress Cataloging-in-Publication Data

Sobel, Dava.
And the sun stood still / Dava Sobel.—First edition.
 pages; cm
"A play in two acts."
ISBN 978-0-8027-1694-1 (hardcover)—ISBN 978-0-8027-7802-4 (ebook)
 1. Copernicus, Nicolaus, 1473-1543—Drama.
 2. Astronomers—Poland—Drama. I. Title.
PS3619.O3738A85 2016
812'.6—dc23
2015027705

2 4 6 8 10 9 7 5 3 1

Typeset by RefineCatch Limited, Bungay, Suffolk
Printed and bound in USA by Berryville Graphics Inc., Berryville, Virginia

To find out more about our authors and books visit www.bloomsbury.com. Here you will find extracts, author interviews, details of forthcoming events and the option to sign up for our newsletters.

Bloomsbury books may be purchased for business or promotional use. For information on bulk purchases please contact Macmillan Corporate and Premium Sales Department at specialmarkets@macmillan.com.

Introduction

MANY MOMENTS in the life of the great astronomer Nicolaus Copernicus, born Niklas Koppernigk in Torun, Poland, in 1473, lend themselves to drama—beginning with the early deaths of his parents, which left him an orphan by age ten. Later the deus-ex-machina intervention of his powerful uncle, a bishop of the Catholic Church, sent him far south to the Jagiellonian University in Krakow, then over the Alps to study law at Bologna, and finally to Padua, where he studied medicine so he could serve as the bishop's personal physician. When he became a churchman himself—a canon of the cathedral at Frauenburg, the City of Our Lady—he undertook arduous and dangerous missions to patrol the landholdings of his diocese. More than once he defended an episcopal palace from marauding bands of Teutonic Knights. In 1522, he headed off a financial crisis by advising the Polish king on how to reform the currency.

Surely one of the most dramatic moments by his own reckoning occurred sometime before 1510, when a

Copernicus holds a lily of the valley, an early Renaissance symbol of a medical doctor (probably because of the flower's association with the god Mercury, whose snake-entwined caduceus promoted healing), in this wood-block portrait by Tobias Stimmer.

phenomenal stroke of insight convinced him that the Sun, and not the Earth, stood at the center of the heavens. Although Copernicus wrote copiously about his heliocentric theory, he never described the events or thoughts that led him to his stunning conclusions: *The Earth moves. It rotates daily on an axis and travels yearly around the Sun, among the other planets.*

The rapidity of these presumed movements required a leap of faith, for Copernicus's Earth did not merely *turn*; it *spun* at a thousand miles an hour, and *hurtled* through its orbit at many times that rate. Such concepts seemed so foolish, so unlikely, so counterintuitive, that Copernicus feared ridicule for even proposing them. Worse, he dreaded being branded as irreverent for upholding notions at odds with numerous passages of Holy Writ.

Nevertheless the idea took hold of him, because only the motion of the Earth could account for the wanderings and relative positions of the other planets. He devoted three decades to the explication of his theory in a lengthy treatise, now known as *On the Revolutions of the Heavenly Spheres*. It was a big book—approximately four hundred folio pages in Copernicus's small, neat hand—dense throughout with mathematical jargon, charts, tables, and geometrical diagrams. He made no effort to publish the manuscript, apparently lacking the courage to do so.

Copernicus's indecision grew into a true dilemma upon the arrival in 1539 of an uninvited visitor. Georg Joachim Rheticus, a twenty-five-year-old mathematical prodigy from Martin Luther's Wittenberg, traveled more than five hundred miles to the northern Polish province of Varmia to seek out the aging Copernicus. His timing was terrible. The bishop of Varmia had recently passed a law banishing all Lutherans from the diocese. This same bishop (a successor to Copernicus's uncle) further tested his powers by threatening to deprive some of

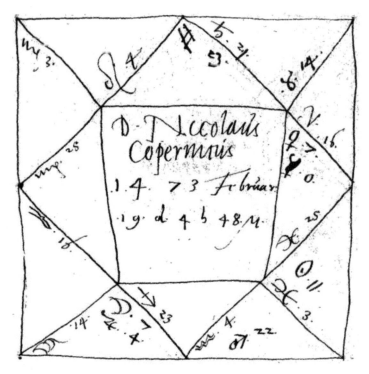

HOROSCOPE FOR NICOLAUS COPERNICUS

Astronomers and astrologers in Copernicus's time shared the same pool of information about the positions of the heavenly bodies against the backdrop of the stars. Until the invention of the telescope in the seventeenth century, position finding and position predicting constituted the entirety of planetary science—and the basis for casting horoscopes.

Copernicus's fellow canons of their sinecures, and also by pressuring Copernicus to drive away his loyal female housekeeper. The last thing Copernicus could reasonably do in this uneasy climate was to harbor a heretic. And yet he agreed to let Rheticus stay with him.

Introduction · ix

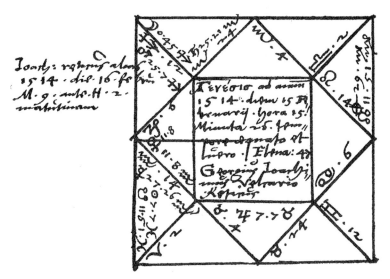

NATAL CHART FOR RHETICUS

The dire prospects suggested in this horoscope for Georg Joachim Rheticus caused his student Nicholas Gugler, who drew the chart, to recalculate the professor's birth time and date. The true date of February 16, written in the margin, disagrees with the more favorable date in the diagram, February 15.

This is the moment I chose to dramatize in my play, *And the Sun Stood Still*. Rheticus's visit is a well-documented historical fact, as is his role in the ultimate publication of Copernicus's book in Nuremberg in 1543. The words that passed between them, however, went unrecorded throughout the two years of Rheticus's sojourn in Varmia. By the strength of his arguments, by the force of his personality, by becoming Copernicus's only disciple, Rheticus convinced his mentor to reverse the reticence of a lifetime.

Other undocumented details in the historical record include the ploys used to hide Rheticus from the bishop's spies, the nature of Copernicus's relationship with his housekeeper, Anna Schilling, and the negotiations that enabled Copernicus to dedicate his magnum opus to the reigning pope, Paul III. All these issues, not to mention the wellspring of Copernicus's convictions, seemed open to fictional exploration.

Copernicus's story first came to my attention in 1973, the five-hundreth anniversary of his birth. The February issue of *Sky & Telescope* featured the only known portrait of Copernicus on its cover, with a capsule biography inside by science historian Edward Rosen. That article introduced me to Rheticus and inspired me to try to write a play, though I did not attempt it right away. I took further inspiration from astronomer Owen Gingerich of Harvard, who devoted many years to locating and examining the several hundred extant copies of *On the Revolutions*, so as to trace the course of its influence. In 2005, the year archaeologists in Poland tore up the floor of Copernicus's cathedral in a search for his mortal remains, I began to imagine the Copernicus-Rheticus exchange.

The version of *And the Sun Stood Still* presented in this volume differs significantly from the one that appeared in 2011 as the centerpiece of my book *A More Perfect Heaven: How Copernicus Revolutionized the Cosmos*. Although the 2011 script served well to humanize its long-dead figures and demonstrate the difficulty of proving the Earth's motion, it was not strong enough to stand up on stage in performance. As I continued revising the dialogue, I sought professional assistance. During a weeklong workshop experience in September 2012 with the Boulder Ensemble Theatre Company in Colorado, at the suggestion of director Stephen Weitz, I eliminated one of the six characters. This change strengthened the remaining relationships and clarified the action. It helped the play make the leap to a well-reviewed world premiere in Boulder in the spring of 2014.

And the Sun Stood Still was originally commissioned by Manhattan Theatre Club, Lynne Meadow, artistic director, Barry Grove, executive producer, with funds provided by the Alfred P. Sloan Foundation. The Solomon R. Guggenheim Foundation also generously supported the project.

And the Sun Stood Still was originally developed and produced by Boulder Ensemble Theatre Company (Stephen Weitz, producing ensemble director), opening March 28, 2014. It was directed by Stephen Weitz, the set design was by Tina Anderson, the costume design was by Katie Horney, the lighting design was by Richard Spomer, and the sound design was by Andrew Metzroth. The cast was:

COPERNICUS	Jim Hunt
BISHOP Johann Dantiscus	Bob Buckley
RHETICUS	Benjamin Bonenfant
ANNA	Crystal Eisele
TIEDEMANN GIESE	Sam Sandoe

AND THE SUN STOOD STILL

A Play in Two Acts

✺ ✺

CAST OF CHARACTERS

COPERNICUS, age 65, physician and canon (church administrator) in Varmia, northern Poland

BISHOP of Varmia, Johann Dantiscus, age 53

RHETICUS, age 25, mathematician from Wittenberg

ANNA, age 45, housekeeper to COPERNICUS

TIEDEMANN GIESE, age 58, Bishop of Kulm (another diocese in northern Poland) and canon of Varmia

The play is set near the cathedral of Varmia in northern Poland, and concerns actual events that took place there between 1539 and 1543, here condensed.

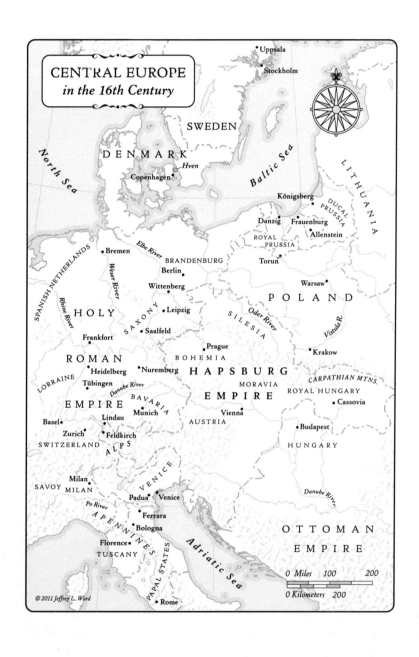

ACT I

Scene i. Bishop's palace

The time is May 1539, in northern Poland.

Darkness. The sound of someone retching. Lights up on Copernicus *standing over the* Bishop, *his patient, who sits on the edge of the bed, vomiting into a basin.*

BISHOP. Oh, God. Oh, Heaven help me.
COPERNICUS. I think that was the last of it, Your Reverence.
BISHOP. Oh, Lord have mercy. Ohhh.
COPERNICUS. The pain will subside now.
BISHOP. I thought I would surely die.
COPERNICUS. The emetic I gave you has rid your body of that toxin.
BISHOP. "Toxin"?
COPERNICUS. It's all gone now. Your Reverend Lordship has expelled it.
BISHOP. Poison?!
COPERNICUS. No, no. I meant . . .
BISHOP. I've been poisoned?!
COPERNICUS. No, Your Reverence. It was . . .

BISHOP. The Lutherans want to kill me.

COPERNICUS. More likely it was something you ate.

BISHOP. Yes, yes. Something I ate. They put it in my food. How else would they get it into me?

COPERNICUS. No. Perhaps just a bite of rotten fish.

BISHOP. The kitchen staff! That shifty-eyed cook.

COPERNICUS. Or maybe too much eel. Your Reverence is extremely fond of eel.

BISHOP. The Lutherans want to assassinate me. Banishing them from the province was not enough to eliminate the threat.

COPERNICUS. Swallow this, Your Reverence. To settle the nerves and bring on sleep.

BISHOP. Sleep?! When Lutheran dogs are trying to kill me?

COPERNICUS. Sleep will be the best thing now.

BISHOP. Worse than dogs. Vermin! Evil and dangerous! Here in our midst, waiting for the moment to strike.

COPERNICUS. Take this, please, Your Lordship.

BISHOP. Suppose they make another attempt on my life, and you don't get here in time? What if . . . ?

COPERNICUS. I must insist that Your Reverence take this medicine now.

BISHOP. I must prosecute them more forcefully. Threaten offenders with harsher punishment. I won't let them get to me the way they got Bishop Ferber.

COPERNICUS. Bishop Ferber?

BISHOP. I see it all now.

COPERNICUS. No one poisoned Bishop Ferber.

BISHOP. They didn't have to! He let them do as they pleased. They walked all over him. Until God Almighty intervened to smite him for not smiting them.

COPERNICUS. Bishop Ferber died of syphilis.

BISHOP. One of God's favorite punishments.

Beat.

BISHOP. Aah! It's done now. He's in his grave, and may he rest in peace. But why did he have to leave the whole Lutheran mess in my hands?

COPERNICUS. Please, Your Reverence.

BISHOP. I must deal harshly with them. I cannot afford to show weakness.

COPERNICUS. You need to rest now.

BISHOP. Pour me a glass of my Moldavian wine. That wine is the better tonic. To strengthen me for the fight. I must issue a new edict.

COPERNICUS. Not now. Not tonight.

COPERNICUS *pours, puts medicine in the* BISHOP's *goblet.*

BISHOP. This time I'll ban their books, too. Ban them and burn them. Build a big bonfire in the town square to consume all those Lutheran Bibles. And their music is anathema. No one may sing those hateful hymns any longer. On pain of . . . of . . .

COPERNICUS. *(handing the goblet to* BISHOP*)* To good health, Your Reverence.

BISHOP. Aaahhh. Here's our spirit.

BISHOP *and* COPERNICUS *drink.*

BISHOP. Agh! Curse that poison! It's killed the taste of pleasure.

COPERNICUS. Rest will restore all. I'll leave Your Reverence to its benefits.

BISHOP. Don't hurry off, Nicholas.

COPERNICUS. Sleep will be the best company now, and will soon arrive.

BISHOP. Stay. Your conversation is a comfort to me.

COPERNICUS. I would not keep Your Reverence from the sleep I have prescribed.

BISHOP. Stay with me a while longer. Your harlot can wait.

COPERNICUS. My . . . ?

BISHOP. Don't give me that innocent look. You know what I'm talking about.

COPERNICUS. If you mean my . . .

BISHOP. That woman. Your harlot! I've told you to send her away. Why is she still in your house?

COPERNICUS. She . . .

BISHOP. She's not Lutheran, is she?

COPERNICUS. No!

BISHOP. Get rid of her anyway. It looks bad. Keeping an unmarried woman like that.

COPERNICUS. She tends to my house. She . . .

BISHOP. She's not even related to you. It's unseemly.

COPERNICUS. The cooking, the . . .

BISHOP. Her manner offends me. She's much too . . . seductive.

COPERNICUS. She's done nothing wrong.

BISHOP. Get yourself an old hag. Or a boy, to take care of your . . . needs. Listen, Nicholas. For myself, I don't care who's in your bed. I understand a man's appetites. God knows, I sowed my oats. Fathered a child or two, here and there, before . . . (*yawns*) But it's different now. With Luther and his fiends screaming to high Heaven and Rome about Church abuses, a man in your position . . . A canon of this Cathedral! You must appear above reproach.

COPERNICUS. Yes, Your Reverence.

BISHOP. Go on home now. Tell her to find a new position. Someplace far away from here.

COPERNICUS sighs, exits.

Blackout.

SCENE ii. OUTSIDE COPERNICUS'S HOUSE

COPERNICUS walks home with a lantern. Near the door of his house, he discovers RHETICUS lying on the ground. COPERNICUS jumps back, then stoops to examine him, checking his pulse, loosening his clothing.

RHETICUS. Ho! Get off me!
COPERNICUS. I was just trying to . . .
RHETICUS. Get away from me!
COPERNICUS. I thought you were . . .
RHETICUS. Thief!

RHETICUS pounces on COPERNICUS. They scuffle.

COPERNICUS. No! Oh!

RHETICUS. What did you take from me?
COPERNICUS. I didn't . . . Oh!
RHETICUS. Give it back!
COPERNICUS. Don't!
RHETICUS. Give it back or I'll strangle you.
COPERNICUS. I'm a doctor.
RHETICUS. What?
COPERNICUS. I am a doctor. I thought you were hurt. I was trying to help you.

RHETICUS releases COPERNICUS, then stands.

COPERNICUS tries to stand.

RHETICUS. Don't move.
COPERNICUS. Who are you?
RHETICUS. You scared me to death.
COPERNICUS. I thought you were dead. I thought . . .
RHETICUS. I was just sitting there, when you came along and . . .
COPERNICUS. You were lying on the ground.
RHETICUS. I was sitting there. Waiting to see . . .
COPERNICUS. Ow!
RHETICUS. Are you hurt?
COPERNICUS. My ankle. I think I . . .
RHETICUS. When you fell, you must have twisted it.
COPERNICUS. Give me a hand, will you?

RHETICUS helps COPERNICUS to sit on a bench in front of the house.

RHETICUS. I'm sorry I hurt you, Doctor. I didn't know . . .

COPERNICUS. Who are you? What are you doing here?

RHETICUS. I'm waiting for Canon Copernicus. Someone told me this was his house. It is his house, isn't it?

COPERNICUS. What do you want with him?

RHETICUS. Is he sick? Is that why you've come?

COPERNICUS. No, he's not sick.

RHETICUS. Thank God. Imagine if I'd come all this way, only to find the great canon, the *starry* canon, too sick to receive me.

COPERNICUS. What did you call him?

RHETICUS. Please forgive me, Doctor. I don't normally get into fist fights. You may not believe this, but I'm a scholar by profession.

COPERNICUS. You?

RHETICUS. A mathematician.

COPERNICUS. Really?

RHETICUS. Professor of mathematics, in fact. My name is Rheticus, Sir. (*extending his hand*) Georg Joachim Rheticus.

COPERNICUS *starts to extend his own hand.*

RHETICUS. (*continued*) Of the mathematics faculty at Wittenberg.

COPERNICUS. (*withdrawing his hand*) Wittenberg?!

RHETICUS. You've heard of it, of course?

COPERNICUS. You came here? From Wittenberg?

RHETICUS. To speak the truth, I was stopping at Nuremberg when I decided to come here.

COPERNICUS. But Wittenberg is . . .

RHETICUS. Nuremberg is even farther away. It added another hundred miles to my journey.

COPERNICUS. But it's not safe.

RHETICUS. Not safe to travel anywhere these days. Between the bandits and the dogs. And the rain! Twice in one day I was almost drowned fording a river.

COPERNICUS. From Wittenberg.

RHETICUS. Canon Copernicus will know its reputation . . .

COPERNICUS. Indeed.

RHETICUS. . . . as a place where the study of mathematics has always flourished.

COPERNICUS. This is Poland, Professor. Catholic Poland.

RHETICUS. I'm certain Canon Copernicus will welcome me, as a fellow mathematician.

COPERNICUS. He will do no such thing. He cannot.

COPERNICUS rises with difficulty.

RHETICUS. Steady there, Doctor.

COPERNICUS. And now, Professor, I bid you a good night.

RHETICUS. Are you sure you can walk?

COPERNICUS. I'm fine. *(indicating the house)* And I don't have far to go.

RHETICUS. Here? But . . . You mean you are . . . ?

COPERNICUS nods assent.

RHETICUS. *(kneeling)* Oh, no! Oh, please forgive me, learned Sir!

COPERNICUS. Now, now. Don't . . .

RHETICUS. All the times I pictured our meeting, and to think . . . Dear Lord, how I've botched things!

COPERNICUS. It's all right. I'll be fine. But you had better move on. This is no place for you.

RHETICUS. If only you knew how I . . .
COPERNICUS. Please, get up.
RHETICUS. The whole way here, I rehearsed, over and over, what I would say when I met you.
COPERNICUS. Say it, then. On your feet.
RHETICUS. (*rising*) Canon Copernicus, I . . . Is that the right way to address you, Sir? Or should I call you "Father"? Did you say you were a doctor?
COPERNICUS. It doesn't matter. Say your piece.
RHETICUS. Begging your pardon, Canon, Doctor, Sir. I have letters here from . . . (*fishing in his satchel*) Letters of introduction from . . .
COPERNICUS. Don't bother with that.
RHETICUS. From Schöner in Nuremberg. And another one from Hartmann, and also Peter Apian, and . . .
COPERNICUS. Did you say, from Schöner?
RHETICUS. Yes, Sir. (*handing him the letter*) Here it is. He was gracious enough to let me stay several weeks with him, in his home. This one is from Camerarius, in Tübingen. He tried to convince me not to look for you. He said you must be dead by now. Excuse me, Canon, Sir. I meant no offense. It's just that no one has heard from you in such a long time. They're all waiting. They wonder why you've kept silent all these years.
COPERNICUS. I have . . . nothing to say.
RHETICUS. You are too modest, Sir. What you've done . . . Why, you have made the greatest leap in astronomy since . . . since Ptolemy introduced the equant. (*brandishing the letters*) Everyone speaks of you. "The starry canon of Poland," they call you, "who

spins the Earth and makes the Sun stand still." They say you've been working at your theory for more years than I've lived.

COPERNICUS. I'm finished with all that now.

RHETICUS. You've finished? You mean you're ready to release the details?

COPERNICUS. There's nothing here for you, Professor. You should go back to Wittenberg. I'm sure your students miss you.

RHETICUS. I am on special leave, on a personal mission, meeting with the most learned mathematicians of our day. I believe that you, Sir, are the culmination of my quest. The very key to the perfection of the heavenly spheres. Doctor, Canon, Sir, I seek to restore the queen of mathematics, that is, Astronomy, to her palace, as she deserves, and to redraw the boundaries of her kingdom.

Beat.

COPERNICUS. I cannot help you.

RHETICUS. Only you can help me.

COPERNICUS. Good night, Professor.

RHETICUS. And I can help you, too.

COPERNICUS. I wish you a safe journey.

RHETICUS. Hear me out!

COPERNICUS. Hush! If the sentry on guard should hear you . . .

RHETICUS. Listen to me.

COPERNICUS. I'm telling you, for your own safety, to leave this place.

RHETICUS. In the middle of the night?

COPERNICUS. It's the best way. So that no one will see you.

RHETICUS. After I've traveled three weeks and five hundred miles just to find you?

COPERNICUS. Be quiet. Come inside. But only for a little while.

Blackout.

SCENE iii. INSIDE COPERNICUS'S HOUSE

COPERNICUS *and* RHETICUS *enter the house.*

COPERNICUS. Now, what were you raving about out there?

RHETICUS. In all the confusion, I never gave you the gifts I brought, Canon, Sir.

COPERNICUS. I could not accept any gifts.

RHETICUS. *(pulling books from his satchel)* You must.

COPERNICUS. Thank you, but no.

RHETICUS. Here is Ptolemy.

COPERNICUS. I have read Ptolemy, of course. Every astronomer has read Ptolemy.

RHETICUS. You read a Latin translation, from the Arabic or the Hebrew. This is the original Greek text.

COPERNICUS. Oh?

RHETICUS. Only recently recovered and now published for the first time.

COPERNICUS. Let me just have a look at that.

RHETICUS. And this is Euclid's Geometry, also in Greek. And here, Regiomontanus, on triangles. I love the part at the beginning, where he says, "No one can bypass the science of triangles, and reach a satisfying knowledge of the stars."

COPERNICUS. These are magnificent volumes.

RHETICUS. I chose the ones I hoped would most please you.

COPERNICUS. You should keep these for your own library.

RHETICUS. They're yours. I've already inscribed them to you.

COPERNICUS. (*reading*) "To N. Copernicus, my teacher..." Your teacher?

RHETICUS. I was hoping...

COPERNICUS. I have no students. No followers of any kind.

RHETICUS. That is why I have come. To be your disciple. Whatever problems have interfered, kept you from bringing your theory to completion, I want to help you solve them. I showed you my letters of reference. Even Melanchthon says I have exceptional aptitude.

COPERNICUS. Philipp Melanchthon?!

RHETICUS. The one they call "teacher of Germany," yes.

COPERNICUS. Luther's own chosen successor? His right hand?!

RHETICUS. He said I was born to study mathematics.

COPERNICUS. Tell me, Professor: Are you on intimate terms with Luther, too?

RHETICUS. Oh! Now I see what you... But I swear to you, Sir, I do not share the Reverend Luther's opinion of your theory. No, indeed.

COPERNICUS. Martin Luther has an opinion about my theory?

RHETICUS. It's only his opinion. Whereas, I feel, astronomy requires precisely the kind of bold new approach that you take.

COPERNICUS. What does he say about it?

RHETICUS. Oh. Things come up at faculty meetings. Lunches. The table talk. You know how it is.

COPERNICUS. No.

RHETICUS. Someone gave him the gist of it, and . . .

COPERNICUS. And?

RHETICUS. Well . . .

COPERNICUS. What did he say?

RHETICUS. He said, anyone who would turn the whole of astronomy upside down, merely for the sake of novelty, must be a fool.
Beat.

COPERNICUS. I suppose "fool" is a mild insult, coming from him.

RHETICUS. And of course he knows nothing of mathematics. He rejected your theory only because it contradicts the Bible. He quoted the Book of Joshua.

COPERNICUS. Ah, yes.

RHETICUS. The part where Joshua says, "Sun, stand thou still upon Gibeon."

COPERNICUS. I know the part, yes.

RHETICUS. "And thou, Moon, in the valley of Aj-a-lon."

COPERNICUS & RHETICUS. (*together*) "And the Sun stood still."

RHETICUS. "And the Moon stayed, until the people . . ."

COPERNICUS. That will do. I am quite familiar with the text.

RHETICUS. Then you see his point, Canon, Sir. The Sun stood still. Whereas, if the Sun were already standing still, as you claim, why would Joshua command it to do so?

COPERNICUS. Why do you think?

RHETICUS. I say, mathematics is for mathematicians. Scripture doesn't enter into it.

COPERNICUS. Is that what you told Luther?

ANNA *enters, weary and disheveled, with blood on her dress.*

ANNA. *(to* RHETICUS*)* You?!
COPERNICUS. Anna.
RHETICUS. She wouldn't let me in when I arrived.
ANNA. Who is this man, Mikoj?
COPERNICUS. What's all this? What has happened to you?
RHETICUS. I thought she was going to throw water on me.
COPERNICUS. *(to* RHETICUS*)* Please.
ANNA. He came to the door just after you left. I was afraid to let him in.
COPERNICUS. You were right to be cautious, my dear. But look at you. Tell me what happened.
ANNA. Oh, Mikoj, the bishop! Is he . . . ?
COPERNICUS. He's all right. Tell me, whose blood is this?
ANNA. The miller's wife. She miscarried again. The poor woman.

COPERNICUS *embraces her.*

RHETICUS. Here they are!
ANNA. Who is he, Mikoj?
COPERNICUS. No one. It's all a mistake. He was about to leave.
RHETICUS. If I could just show you these . . .
COPERNICUS. Not now.
ANNA. What does he want from you?
COPERNICUS. Forget about him. He'll be gone in a moment. You'll see. Let's get you cleaned up and into bed.
ANNA. But why . . . ?

COPERNICUS. Please, my dear. Trust me. It's better that you don't know.

ANNA. I don't like the look of him.

COPERNICUS. (*bundling her off to bed*) I'll take care of him. Don't worry yourself.

ANNA. Be careful, Mikoj.

ANNA exits.

RHETICUS. Now, then. Wait till you see . . .

COPERNICUS. I hate to be inhospitable, Professor. But now you really must . . .

RHETICUS. Look what else I brought you, Sir.

COPERNICUS. No, please. No more gifts.

RHETICUS. These notes are from Schöner. Some recent observations he collected, of Mercury. He insisted that I bring them to you.

COPERNICUS. (*taking the diagrams*) He did?

RHETICUS. He didn't make the observations himself. He said they were the work of some other astronomer. But he said they would please you.

COPERNICUS *shakes his head with wonder, nods in admiration, sighs.*

RHETICUS. You haven't given up, have you, Sir? Am I right? Canon, Sir?

COPERNICUS. Hm?

RHETICUS. I said, you haven't quit. Have you? It's just taking time. Isn't that right? That's why I thought I could help.

COPERNICUS. No. I'm sorry. Even if I wanted to, I . . . My hands are tied. The bishop, you see, has . . . He . . . I fear there's no nice way to say this, Professor. The bishop has banished all Lutherans from this diocese.

RHETICUS. What has that got to do with me?

COPERNICUS. You mean you're not? Lutheran?

RHETICUS. I'm not looking to settle down here. I merely wish to talk to you, about your work.

COPERNICUS. Even that would not be allowed. No.

RHETICUS. I'm a mathematician, not a theologian.

COPERNICUS. Still.

RHETICUS. Couldn't you explain that to him? Perhaps he would grant us a . . . What do you call it? An indulgence?

COPERNICUS. A dispensation. But, no. There's no chance of that.

RHETICUS. Oh, please try. You know yourself that our discussions need never touch on faith.

COPERNICUS. No, even to discuss . . .

RHETICUS. We'll limit ourselves strictly to arithmetic and geometry. The wings of the human mind. On such wings as those, we transcend our religious differences. Transcend all religious differences. Didn't Abraham teach astronomy to the Hebrews? And Moses, another Jew? And Heaven knows how all the Islamic astronomers prayed to their Allah five times a day, then watched the stars all night. Even going back to the Egyptians, the Greeks! Prometheus and the theft of divine fire! The very crime for which he suffered an eagle to devour his liver! What does that mean, if not that Prometheus delivered the light of astronomy to all mortals?

Beat.

COPERNICUS. How young you are, Professor.

RHETICUS. You're not afraid to talk to him, are you?

COPERNICUS. Afraid? I am the bishop's personal physician.

RHETICUS. Well, then.

COPERNICUS. I was summoned to his side tonight, after he was poisoned by a Lutheran spy.

RHETICUS. No!

COPERNICUS. No. It was . . . nothing like that. But knowing the intimate details of his digestion does not give me leverage to sway his opinion. On any subject.

RHETICUS. *(kneeling)* Please try! I implore you. If you do, I swear I will . . .

COPERNICUS. Come, Professor. You must leave off this genuflecting and swearing. Remember, you are not a Catholic, and I am not a priest.

RHETICUS. You're not?

COPERNICUS. Only minor orders. But I administer the Cathedral's business affairs. I'm an officer of the Church. Don't you understand? I cannot harbor a heretic.

Beat.

COPERNICUS. *(continued)* I'm sorry if I've offended you. I meant no disrespect for your beliefs.

RHETICUS. You mean . . . I'd be a danger to you?

COPERNICUS. You are a danger to yourself, young man. Rushing off to unknown places, knocking on strangers' doors, shouting about missions and quests.

RHETICUS. I only wished to . . .

COPERNICUS. Now, Professor. Gather your things. I hate to send you away like this. But we are victims of these times.

RHETICUS. You mean, you won't let me stay?

COPERNICUS. I'm sorry. And take the books, please. I could not keep them in good conscience.

RHETICUS. What will I do now? How will I ever . . .? Oh, please, reconsider!

COPERNICUS. Why don't you write to me? After you get back to Wittenberg, you could . . . Not write to me directly, of course. You would need to send your letters through an intermediary. Someone who could serve as a point of contact for us. (*putting a friendly arm around* RHETICUS) Come now, Professor. Pull yourself together. And be cautious on the roads. Mind you take care of yourself out there.

Together they walk to the door. When COPERNICUS *opens it, daylight floods the room.*

COPERNICUS. Oh, for Heaven's sake!

COPERNICUS *shuts the door and pushes* RHETICUS *back into the room.*

COPERNICUS. (*continued*) You can't go now!

RHETICUS. Sir?

COPERNICUS. It's too late. Daylight already. I'll have to . . . Where . . . ? I know!

COPERNICUS *pushes against a bookcase that gives way to a secret passage.*

COPERNICUS. Now come this way.
RHETICUS. What?
COPERNICUS. Hurry.

They exit through the secret door.

Blackout.

SCENE iv. BISHOP'S PALACE

BISHOP *sleeps.*

A knock at the door frightens him, and he cries out.

TIEDEMANN GIESE, *the bishop of Kulm, enters.*

GIESE. May I come in?
BISHOP. No!
GIESE. Were you sleeping, Johann?
BISHOP. No, of course not.
GIESE. Is anything the matter?
BISHOP. What are you doing here, Tiedemann?
GIESE. I came to beat you at chess, as usual. Have you forgotten?
BISHOP. Stop shouting!
GIESE. I'm not shouting. What's wrong with you, Johann? You look terrible.
BISHOP. I feel terrible. Like a horse kicked me in the head.
GIESE. We should send for Nicholas to come and examine you.
BISHOP. Nicholas was here all night. An awful night I had. Cursed Lutherans tried to poison me.

GIESE. Poison?!

BISHOP. They tried to kill me. And very nearly succeeded.

GIESE. Heaven forbid.

BISHOP. Agh! I ask you, Tiedemann: If I'm not safe in my own dining room, where am I safe? Lutherans everywhere. In the kitchen. In the soup.

GIESE. But you seem well enough now. Are you sure it was poison?

BISHOP. Do you doubt my word?

GIESE. What did Nicholas say?

BISHOP. Nicholas! His skills may combat a single instance of poisoning. And thank God for that. But his medicaments cannot stanch the spread of the Lutheran menace. We must fight harder to contain its foul contagion. As God is my witness, we must take up arms against it!

GIESE. You talk like a soldier, Johann.

BISHOP. And you, Tiedemann! You sit idly by, and watch. You do nothing to stem the tide.

GIESE. What would you have me do? Lay siege to Wittenberg?

BISHOP. All I ask is that you adopt my edict in your diocese.

GIESE. No, Johann. That I cannot do. I see no need to put such a harsh ruling into effect.

BISHOP. After last night, I intend to strengthen it. Increase the prohibitions. Magnify the punishments.

GIESE. Oh, no, Johann.

BISHOP. We're the only ones left, Tiedemann. You and I. We're the last holdouts in the whole region. Every other bishop, to a man, has bowed to that devil Luther. God help us, even the duke has converted. We are surrounded. We must crush them.

GIESE. We are men of God, Johann.

BISHOP. The Church calls us to her defense. I need your support. How can you continue to allow Lutherans to live and work in Kulm?

GIESE. Our Lutherans in Kulm don't cause trouble. They just go about their . . .

BISHOP. Listen to me, Tiedemann. If we have trouble here in Varmia, you have trouble in Kulm. We have the same troubles, you and I. How do you know my assassin wasn't one of your Lutherans?

GIESE. These are peasant farmers. Merchants. Tradesmen. The same people who have lived among us for generations, since long before . . .

BISHOP. They have betrayed us, by betraying the Church. You must expel them.

GIESE. In your heart, you know there's a better path to reconciliation with our Protestant brethren.

BISHOP. Oh, please! When will you face the facts?!

GIESE. We're all Christians in the eyes of God.

BISHOP. Haven't the past twenty years taught you anything? That sniveling little monk! He has whined and complained and . . . and gained himself a huge following! How did it happen? Hm? Who ever thought anyone would listen to him? Now look at him. He sings a few hymns, and half the continent thinks he's the Second Coming. It's an abomination.

GIESE. The Church has weathered worse storms before this. If we are steadfast in our faith, and treat our fellow citizens with compassion . . .

BISHOP. You mean you refuse to back me?

GIESE. I'm saying that the changing times challenge us to summon new reserves of patience and understanding.

BISHOP. Patience! What do you think this is? A game of chess?

GIESE. Let us pray together, for guidance. "Pater noster, qui es in caelis . . ."

BISHOP. I bet you'd just love for one of them to do away with me.

GIESE. Don't give in to your dark thoughts, Johann. Pray with me now. "Pater noster, qui es in caelis, sanctificetur nomen tuum . . ."

BISHOP. Then you could take my place, and be bishop here yourself.

GIESE. "Adveniat regnum tuum. Fiat voluntas tua, sicut in caelo et in terra."

BISHOP. That's why you keep your canonry here, isn't it?

GIESE. "Panum nostrum quotidianum da nobis hodie."

BISHOP. You want to have your foot in the door, so when I die . . .

GIESE. "Et dimitte nobis debita nostra, sicut et nos dimittimus debitoribus nostrus."

BISHOP. Why didn't I see it before?

GIESE. "Et ne nos inducas in tentationum, sed libera nos a malo."

BISHOP. Why else would you remain a canon here in Varmia?

GIESE. "Quia tuum est regnum, et potestas, et Gloria, in saecula. Amen."

BISHOP. You should give up your canonry!

GIESE. What?!

BISHOP. You have no right to be a canon here any longer.
GIESE. Don't be ridiculous, Johann. I have every right . . .
BISHOP. I want you to resign. Do it now. Step down of your own volition. Don't make me force you out.
GIESE. You cannot force me. A canonry is a lifetime appointment. As you well know.
BISHOP. Nevertheless, you are free to leave it if you choose.
GIESE. I would never do that. I rely on my income from the canonry.
BISHOP. You're bishop of Kulm now.
GIESE. Kulm is a poor diocese. You know that better than anyone, Johann. When you were bishop of Kulm, you . . .
BISHOP. You cannot be bishop of Kulm and canon of Varmia, too.
GIESE. Of course I can. You did. When you were made bishop of Kulm, you held on to your Varmia canonry.
BISHOP. What I did has nothing to do with what you should do.
GIESE. It is exactly the same situation. You remained a canon here the whole time you were bishop of Kulm. If you hadn't done that, you could never have been elected bishop of Varmia.
BISHOP. Aha! You admit it then! You do want to take my place!
GIESE. I'm older than you, Johann. I'm not likely to outlive you.
BISHOP. Not likely, no. Except in the event of my untimely death.
GIESE. You cannot accuse me of such treachery!
BISHOP. Can't I?
GIESE. It's the principle of the thing. And the income, of course. And I . . . I still belong to this community. These are my lifelong friends. Why, Nicholas and I go back . . .

BISHOP. Don't expect your friend Nicholas to come to your rescue now. He's on very shaky ground himself.

GIESE. Nicholas?! Nicholas keeps all of us alive!

BISHOP. I could see him excommunicated.

GIESE. Have you gone mad, Johann?

BISHOP. I refuse to look the other way any longer while that harlot comes and goes as she pleases.

GIESE. You mean the housekeeper?

BISHOP. Housekeeper, harlot. What's the difference? What do you take me for? A simpleton?

GIESE. At least three or four of the other canons keep female housekeepers. It's always been that way.

BISHOP. Not any more. Not since I've taken up the reins here. All those women have been discharged, all except for her. She's made a cuckold of him, too, sneaking out at night, behind his back. She may be in league with the Devil.

GIESE. Stop it, Johann. You don't know what you're saying.

BISHOP. You put a stop to it! Tell him to get that woman out of here. That's the least you could do in support of my wishes. Go and reason with him. Let him hear it from both of us. If you want to keep your canonry, then show me your value here in Varmia.

Blackout.

Scene v. Tower Room

Dim lights reveal the Tower Room as small, spare, and dusty from disuse. The furnishings include a table and chair, a cot, and the World Machine, a globe-like nest of intersecting rings, about the size of a manned spacecraft capsule.

RHETICUS. (*offstage*) Where are you taking me, Sir?

COPERNICUS. (*offstage*) Only a little farther now.

RHETICUS. (*offstage*) But where . . . ?

COPERNICUS. (*offstage*) We're nearly there . . . (*entering*) Ah! Here we are. You can stay here.

RHETICUS. Here? What is this place?

COPERNICUS. You'll be safe here.

RHETICUS. Is it your observatory?

COPERNICUS. This? No.

RHETICUS. Not a prison cell, is it?

COPERNICUS. Oh, no. It's a retreat. A safe place. We all have rooms like this. When there's danger of enemy attack, we come here, and we stay, until it's safe for us to leave.

RHETICUS. You really expect me to stay here?

COPERNICUS. Just till tonight. After sunset, you can go. As soon as it's dark, I'll come fetch you.

RHETICUS. You mean you're not staying with me?

COPERNICUS. No.

RHETICUS. But now we have all day.

COPERNICUS. I can't stay with you. I have duties I must . . .

RHETICUS. Oh, please. Stay and seize this day with me. Look

how God has provided a space of time for us, after all. This is our chance to speak as one mathematician to another. In these secluded surroundings, you and I could . . .

*R*HETICUS *sees the Machine.*

RHETICUS. *(continued)* What is that?

COPERNICUS. That? That's just . . . something I made.

RHETICUS. You built it?

COPERNICUS. A long time ago.

RHETICUS. But what is it? Some kind of observing instrument?

COPERNICUS. No. No, it's . . . more of a model, really.

RHETICUS. Like an armillary sphere?

COPERNICUS. Something like that.

RHETICUS. Only larger.

COPERNICUS. Yes.

RHETICUS. Much larger.

COPERNICUS. I don't use it any more.

RHETICUS. Why so big?

COPERNICUS. Well, the person inside needs room to . . .

RHETICUS. There's someone inside it?!

COPERNICUS. Not now.

RHETICUS. No. But a person could . . . ?

COPERNICUS. Yes. The person has to sit inside it, to get the effect.

RHETICUS. And what effect would that be? Inside?

COPERNICUS. The sense of . . . the consequences, really, of my theory.

RHETICUS. Oh, I see. You sat in there, while you were figuring out how to . . . ?

COPERNICUS. No. I stood out here, to operate it.
RHETICUS. Someone else was inside?
COPERNICUS. Yes.
RHETICUS. So you did have a student? Before me?
COPERNICUS. No.
RHETICUS. Then who . . . ?
COPERNICUS. No, I made this for my . . . for a friend, who couldn't grasp the mathematical concepts. Someone who needed a tangible way to . . . visualize the spheres. She . . .
RHETICUS. You certainly went to a lot of trouble.
COPERNICUS. I suppose I did.
RHETICUS. For your friend.
COPERNICUS. Yes. Well, then. You wait here, and . . .
RHETICUS. Could I try it?
COPERNICUS. No, I don't think . . .
RHETICUS. I'd really like to see what it does.
COPERNICUS. No one's used it in years. I doubt it still works.
RHETICUS. Let's try it and see.
COPERNICUS. There's no need for that. You, of all people, can follow the math.
RHETICUS. I was hoping to read your work, Sir. I didn't know I could ride in it.
COPERNICUS. Don't touch that.
RHETICUS. How do you get in?
COPERNICUS. Not there. No, not like that.
RHETICUS. Show me, then. Please.
COPERNICUS. Let go of that. It's over here. You climb in through here.

> RHETICUS *dives in, but finds entry a struggle.*

RHETICUS. Your friend must have been quite small. There's hardly room to . . .

COPERNICUS. Maybe you shouldn't . . .

RHETICUS. All right. I'm in. It's very dark in here. I don't see anything.

> *Dim lights come up, just enough to show* RHETICUS *sitting inside the Machine.*

COPERNICUS. I'm lighting it now . . . Give me a moment to . . .

RHETICUS. What?

COPERNICUS. I said I'm lighting it now.
There!

> *Little twinkling star lights appear, as in a planetarium.*

COPERNICUS. Do you see anything?

RHETICUS. Good Lord!

COPERNICUS. You see?

RHETICUS. Good heavens. It's . . . There are stars everywhere. All around. How did you do that?

COPERNICUS. Now I turn you.

RHETICUS. What?

COPERNICUS. I said, I'll turn you around now.

RHETICUS. Should I do something?

COPERNICUS. No . . . Ugh . . . Ah, there it goes!

   RHETICUS *starts rotating in his seat.*

RHETICUS. Good God! What's happening? Oh, this is . . . This is unbelievable.

COPERNICUS. You see? You see what it does?

RHETICUS. Oh, Sir! You have reproduced the night. The effect is . . . It's spectacular.

COPERNICUS. Now you see how the device approximates the motions.

RHETICUS. It's so lovely. So faithful to the zodiac constellations. I've never seen anything to compare with this. Look! There's the ram, the bull, the twins, the . . .

COPERNICUS. I'm afraid I can't keep this up much longer. Not as strong as I used to be.

   RHETICUS's *seat slows to a stop. The stage lights return.*

COPERNICUS. You'd better come out now.

RHETICUS. Oh, my. That was . . .

COPERNICUS. Steady, there.

RHETICUS. I'm still seeing stars.

COPERNICUS. Let your eyes adjust to the light.

RHETICUS. Thank you, Canon, Sir.

COPERNICUS. Did you find it convincing?

RHETICUS. Convincing?

COPERNICUS. Did you?

RHETICUS. Convincing of what, Sir?

COPERNICUS. Of the motion.

RHETICUS. Oh, most definitely.

COPERNICUS. Good. Well, then.

RHETICUS. All the stars moved. I could see them spinning round and round.

COPERNICUS. No, the stars didn't . . .

RHETICUS. It was wonderful.

COPERNICUS. That was you going around. Not the stars.

RHETICUS. No, I saw the . . . The stars turned around me.

COPERNICUS. You turned. In that little seat. That is the only part that moves.

RHETICUS. But I didn't feel myself move.

COPERNICUS. You're not supposed to.

RHETICUS. I am quite certain that I . . . that the seat did not move.

COPERNICUS. That's just it. That's the point, you see. You think the stars are turning, when really it's you turning. And once you realize that you are the one going around, then you make that shift in perception. You see?

RHETICUS. I'm not sure I do. No.

COPERNICUS. The Machine gives you a physical appreciation. For the way the Earth, by its rotation, makes the stars appear to spin around it. And the planets, too. I tried to build in the planetary effects . . . the stations and retrogrades . . . but I had trouble aligning them.

RHETICUS. Do you mean to say . . . ?

COPERNICUS. I think those parts must still be around here, somewhere . . .

RHETICUS. *(lurching, a little dizzy)* Oh, no!

COPERNICUS. What?

RHETICUS. You mean, you really do mean to turn the Earth?
COPERNICUS. You knew that.
RHETICUS. But . . . really turn it?
COPERNICUS. What did you think?
RHETICUS. I didn't think you meant to turn it . . . physically.
COPERNICUS. How else would it turn, if not physically?
RHETICUS. It would turn . . . theoretically. You know. In a hypothetical way. On paper. In order to . . .
COPERNICUS. No.
RHETICUS. Theoretically. Mathematically. But not . . .
COPERNICUS. No, the motion is real. Of course it is.
RHETICUS. How could you claim such a thing?
COPERNICUS. I thought you understood what my work was about.
RHETICUS. (*another rush of dizziness*) I . . .
COPERNICUS. Didn't Schöner explain it to you?
RHETICUS. He, uh . . . I . . .
COPERNICUS. What did he tell you?
RHETICUS. I don't think he sees it quite the way you do, Sir.
COPERNICUS. How can that . . . ?
RHETICUS. He didn't mention anything about a real motion.
COPERNICUS. Are you sure?
RHETICUS. All he said was . . . No, he didn't say anything about . . .
COPERNICUS. You mean, he doesn't understand it either?!
RHETICUS. I think he must not have interpreted it . . . literally.
Beat.
RHETICUS. (*continued*) Why would he?

COPERNICUS. Why?!
RHETICUS. Why would he leap to that conclusion?
COPERNICUS. Oh, merciful heavens!
RHETICUS. Honestly, Sir. I don't think anyone realizes exactly what it is that you have in mind.
COPERNICUS. What can they think I've been doing all these years?
RHETICUS. Even just to . . . to use the idea as the basis for new calculations, would . . . But, to claim the motion as reality?!
COPERNICUS. Yes.
RHETICUS. I am . . . I . . . Look! You and I. We're just standing here. The Earth . . . (*wobbling, but stomping his foot for emphasis*) It doesn't move.
Beat.
COPERNICUS. Yes it does.
RHETICUS. You really believe the Earth is . . . turning?
COPERNICUS. This is not a question of belief, Professor. I know it turns.
RHETICUS. What do you mean, you "know"?
COPERNICUS. I mean the evidence has convinced me.

A peal of bells is heard. It continues through the end of this scene.

RHETICUS. What evidence?
COPERNICUS. Goodness, the time!
RHETICUS. You mean the Earth leaves some kind of wake behind it? Like a boat?
COPERNICUS. I'm sorry. I must leave you now.

RHETICUS. No, wait a minute.
COPERNICUS. You must excuse me. I'll come back tonight.
RHETICUS. Wait!
COPERNICUS. They're expecting me in the . . .
RHETICUS. Just because I raise a few questions? You walk away?
COPERNICUS. Don't you hear the bells? That's the call to Mass.
RHETICUS. Wait! Please!
COPERNICUS. I must get to the Cathedral. If I'm not seen at Mass, it will raise too many questions.
RHETICUS. Wait!

>   RHETICUS *takes a step toward the door, then falls to the ground.*
>
>   *Blackout.*

SCENE vi. INSIDE COPERNICUS'S HOUSE

>   ANNA *has just let* GIESE *into the house.*

GIESE. Not here?
ANNA. No, Your Reverence.
GIESE. I was hoping to meet him, at Mass.
ANNA. He wasn't at Mass?
GIESE. No. I thought he might be ill or . . .
ANNA. Oh, no. Where could he be?
GIESE. You don't know where he is?

Beat.

ANNA. Your Reverence, may I confide in you?

GIESE. You wish to make a confession?

ANNA. As I know Your Reverence to be a true and loyal friend, I feel I must mention something happening in this house.

GIESE. Oh. I think I would rather discuss this with him privately.

ANNA. It's just that last night . . .

GIESE. Yes, yes, I know.

ANNA. How could that be? I thought no one knew about the . . .

GIESE. Please understand. I have nothing against you, Daughter. You mustn't take all the blame on yourself.

ANNA. Me? I didn't let him in. I tried to . . .

GIESE. It's never one-sided in these situations. To be frank, I feel partly responsible. I've known about it all along. And yet I said nothing. As his friend, I should have counseled him. I could have saved him from this . . . this . . . regrettable situation.

ANNA. Is he in some kind of trouble?

GIESE. Don't fret now. Nothing bad will happen to him if you are brave and do what's required of you. Tell me, do you have family who could take you in?

ANNA. What?

GIESE. Or a friend, perhaps? Someplace where you know people, where you could go and feel welcome? And make a new home for yourself?

A scuffling sound comes from behind the secret passage.

COPERNICUS. *(offstage) (panting, whispering)* Anna?

ANNA. Oh!

GIESE. What was that?
COPERNICUS. (*offstage*) Tiedemann? Is that you?
GIESE. Good heavens! Nicholas?
COPERNICUS. (*offstage*) Let me in.

GIESE and ANNA push open the door.

COPERNICUS drags in the unconscious body of RHETICUS.

GIESE. What in the world . . . ?
COPERNICUS. Help me.

The three of them pull RHETICUS into the room.

ANNA. Oh, dear. He's burning up with fever.
GIESE. Who is this?
ANNA. I'll get some blankets.
COPERNICUS. And willow bark.
ANNA. Of course.

              ANNA *exits.*

GIESE. Who is he?
COPERNICUS. He . . .
GIESE. What's wrong with him?
COPERNICUS. He . . .
GIESE. Oh, never mind. You can tell me after you catch your breath.
COPERNICUS. (*embracing* GIESE) Tiedemann.

ANNA returns with blankets, water, and medicine.

ANNA. His clothes are soaking wet.

COPERNICUS. Better get them off.

ANNA. Why did you let him in the house?

GIESE. Who is he?

ANNA. The last stranger you allowed in brought malaria with him.

COPERNICUS. This is nothing but ague. Exposure. Heaven only knows where he's slept in weeks of travel.

GIESE. Where did he come from?

ANNA. That's what I'm saying. You don't know where he's been.

COPERNICUS. But I do, Anna. I know where he . . .

GIESE. You know him, Nicholas?

COPERNICUS. Not really, no.

GIESE. I don't understand.

COPERNICUS. When I came home from the bishop's apartments late last night, I found him, lying in front of my house.

GIESE. You carried him into the house?

COPERNICUS. No, Tiedemann. He walked into the house.

ANNA. He can't stay here.

COPERNICUS. You're right. Better make up a bed for him in the pantry.

ANNA. No, he cannot stay here.

COPERNICUS. Please, Anna. There's nothing else to be done till his fever comes down.

ANNA *exits.*

COPERNICUS. Of all the times for someone like him to . . . Someone of his talents . . . Why now? Agh! If only he'd come to me twenty years ago.

GIESE. Twenty years ago he was still in swaddling clothes, from the look of him.

COPERNICUS. It wouldn't have made a difference then either. My theory was never meant to see the light of day.

GIESE. He came to you about that?

COPERNICUS. So he said.

GIESE. What about it? What did he say?

COPERNICUS. Nothing. It doesn't matter. He didn't really understand it anyway.

GIESE. But who is he?

COPERNICUS. You wouldn't believe me if I told you.

GIESE. Where did he come from, Nicholas? Why were you hiding him?

COPERNICUS. He came here from . . . He brought letters from . . . He had a letter of introduction from Schöner. One from Hartmann, too. And a stack of books he tried to give me. Ptolemy in the original Greek. Can you imagine? And here. Look at these.

GIESE. It's been ages since I looked at diagrams like these. What am I to make of them?

COPERNICUS. They concern the circles of Mercury. All the times I tried, I never could capture Mercury at that large an angle of western elongation. Without such observations, it was impossible for me to complete my own assessment of the anomaly.

GIESE. And this was his gift to you?

ANNA returns.

ANNA. It's ready.

All three pick up RHETICUS *and carry him toward the pantry.*

GIESE *returns alone, thoughtful.*

COPERNICUS *returns.*

GIESE. There is something strange and wonderful about the arrival of this fellow. Don't you think so?
COPERNICUS. In a few days he'll be well enough to return to Wittenberg.
GIESE. To Wittenberg, is it?
COPERNICUS. I told you you wouldn't believe me.
GIESE. Oh, Nicholas. Do you realize what is happening here?
COPERNICUS. I know. I'll send him on his way as soon as . . .
GIESE. No. He should stay.
COPERNICUS. What?
GIESE. Don't you see? He has come to help you. And now he shall be of help to me as well.
COPERNICUS. He is in no condition to help anyone.

ANNA *returns.*

GIESE. I must speak with the bishop.
COPERNICUS. There's no need to tell him about my . . . my patient.
GIESE. One thing is certain, Nicholas. The Lord surely works in mysterious ways.

GIESE *exits.*

ANNA. Why did he want to know if I had family?

COPERNICUS. What?

ANNA. Just now. Just before you . . . He asked did I have family or friends I could go to? What did he mean, Mikoj?

COPERNICUS. The bishop must have told him.

ANNA. Told him what? Oh, Mikoj! The bishop wants you to send me away.

COPERNICUS. (*embracing her*) Don't worry. I would never do that.

ANNA. What if he makes you? Forces you?

COPERNICUS. He'll forget about us.

ANNA. And if he doesn't?

COPERNICUS. I won't let anything happen to you.

ANNA. I see the way he looks at me.

COPERNICUS. Has he dared to lay a hand on you?

ANNA. That lecherous dog of a false priest.

COPERNICUS. But has he? Touched you?

ANNA. I won't go, Mikoj.

COPERNICUS. I won't let you go.

Blackout.

SCENE vii. BISHOP'S PALACE

BISHOP *sits at the desk where he signs and seals the edict.*

BISHOP. He just took him in? A total stranger?

GIESE. Out of compassion, yes.

BISHOP. Without knowing his identity?

GIESE. He found the fellow lying prostrate in front of his house. Stricken down by illness. You know Nicholas. He could not abandon someone in that condition.

BISHOP. For all he knows, the stranger might be a spy.

GIESE. No, he's a mathematician.

BISHOP. I thought you said no one knew anything about him.

GIESE. Well ... He had several mathematics textbooks in his satchel.

BISHOP. Books in a bag don't prove a person's profession.

GIESE. Large, important texts.

BISHOP. Perhaps he stole them.

GIESE. I think he has come here with a purpose, Johann. Expressly to question Nicholas about his theory.

BISHOP. You're making a great number of assumptions.

GIESE. To shake Nicholas out of his long paralysis.

BISHOP. Don't run away with yourself, Tiedemann. We have important matters to address. I want you to sign the new edict. Right there, below where I've affixed my seal.

GIESE. Think what it could mean, Johann. I've always said, someday Nicholas will bring glory to Varmia through his mathematical work.

BISHOP. That is one hare-brained idea, that theory of his. I thought he was wise to put it aside.

GIESE. He should be encouraged to take it up again. If I've tried once, I've tried a thousand times to convince him to see it through.

BISHOP. He should let it lie. It's a dangerous notion.

GIESE. It's controversial, I grant you, but ...

BISHOP. It may even be heretical.

GIESE. Oh, no. Johann.

BISHOP. It makes him a laughingstock, at the very least. You should hear what they say about him at court. How he mistakes the Earth for a side of beef. So he puts it on a spit, and roasts it in the Sun's fire.

GIESE. His ideas are beyond the comprehension of ordinary minds like yours and mine.

BISHOP. Even mathematicians have common sense, Tiedemann. Now, then. Stop changing the subject, and sign. Will you do that? Will you stand with me to protect Varmia? And Kulm. And the rest of our province, and Poland, and the world?

GIESE. I'm sorry, Johann. I cannot condone the punishment of innocent people.

BISHOP. I have already written my recommendation to the provost of the Chapter, requesting that you be relieved of your canonry. I have it right here.

GIESE. I'll not be intimidated by such threats. I cannot believe the Chapter will approve such an action against me.

BISHOP. You sign the edict, Tiedemann, and I'll tear up the letter.

Beat.

GIESE. I must be getting back to Kulm now.

BISHOP. Sign, damn it!

GIESE. I have many preparations to make. I'm planning to invite this newly arrived mathematician to Kulm, as soon as he is well enough to travel.

BISHOP. The sooner he leaves here, the better.

GIESE. And the nurse. To look after him.

Beat.

BISHOP. Why, that's brilliant, Tiedemann! So you did find a way to get her out of Varmia. Ha HA! Well done, and good riddance to her!

GIESE. Naturally I'll have Nicholas come, too.

BISHOP. Nicholas?

GIESE. Yes.

BISHOP. Nicholas isn't going anywhere.

GIESE. He will leap at the chance to engage the visitor in learned discourse. It's the perfect opportunity for Nicholas to resume his great work.

BISHOP. You can't take Nicholas that far away.

GIESE. It's only for a little while. To smooth out the . . . the relocation of the woman.

BISHOP. Nicholas is barred from travel to any diocese where Lutherans are allowed to run free.

BISHOP picks up the edict, thrusts it at GIESE.

GIESE. You're making this impossible, Johann. If Nicholas won't be coming, there's no point in my taking the stranger with me. Or the nurse.

Blackout.

Scene viii. Inside Copernicus's house

Late that night, by candlelight, Copernicus *and* Anna *kneel at an old trunk, remove stacks of pages from it.*

Anna. If the bishop finds out that you are sheltering a Lutheran . . .
Copernicus. I'm housing an invalid. Ministering to the sick.
Anna. If he finds out, he'll banish you both.
Copernicus. Please, Anna. Keep looking. You'll recognize Mercury when you get to it.
Anna. Yes, of course. But why do you need it now? We haven't looked at these pages in such a long time.
Copernicus. Something that young professor brought me inspired a new idea.
Anna. Oh? What are you thinking?
Copernicus. Remember how we could never . . .

Rheticus *staggers in, wrapped in a blanket.*

Rheticus. What happened?
Anna. Oh!
Copernicus. You startled us.
Rheticus. Why didn't you tell me?

Rheticus *stumbles, starts to fall.*

Copernicus *catches him.*

Copernicus. You shouldn't be wandering about.
Rheticus. It's dark now. It's dark!

COPERNICUS *and* ANNA *sit him down.*

RHETICUS. (*continued*) You promised you'd tell me when it got dark.

COPERNICUS. You're ill. Do you remember? Anna, bring some of that broth.

ANNA *exits.*

RHETICUS. Where are my clothes?

COPERNICUS. (*taking off his cassock, putting it around* RHETICUS) You won't be going anywhere tonight. You're still weak. In another day or two, you'll be stronger. Then you can do as you please. But for now you are in my care.

RHETICUS. This is your house. We were in this room.

ANNA *returns with the broth.*

COPERNICUS. Here, drink some of this.

RHETICUS. But this isn't where we . . . We went somewhere else to . . .

COPERNICUS. Go on, drink. It's good for you.

RHETICUS. (*taking the cup of broth*) You put me in that . . . machine.

ANNA. The Machine?

RHETICUS. Oh, NO!

ANNA. You showed him the Machine?

COPERNICUS. (*to* ANNA) It's all right.

RHETICUS. (*dropping the cup*) Now I remember. Oh, no. Oh, no, no, no.

ANNA *exits.*

COPERNICUS. You must have been dreaming. Anna!
RHETICUS. I thought you would save me.
COPERNICUS. Sometimes fever causes very vivid, frightening dreams.
RHETICUS. I thought you could help me. Now what will I do?
COPERNICUS. You'll be fine.
RHETICUS. What will become of me?
COPERNICUS. Good as new, you'll see.
RHETICUS. I came here in good faith . . .
COPERNICUS. Yes, yes. I know.
RHETICUS. And what do I find? A lunatic! A deluded old . . . A . . . A recluse! Obsessed with an insane idea.
COPERNICUS. Get hold of yourself.
RHETICUS. I must go. Where are my clothes? Where's my satchel?
COPERNICUS. You don't need any of that now.
RHETICUS. My horoscopes are in there.
COPERNICUS. I know how to treat your symptoms without looking at a horoscope.
RHETICUS. You don't understand. Where is that satchel?
COPERNICUS. Calm yourself.
RHETICUS. (*despairing*) I can recite it all by heart.
COPERNICUS. What can you recite?
RHETICUS. Every house, every aspect, every conjunction and opposition. Every indicator of doom.
COPERNICUS. Don't tell me you believe in that . . . that . . . ?
RHETICUS. It's not as though I have a choice.
COPERNICUS. You should know better.
RHETICUS. If only I could forget what I know.

COPERNICUS. Why don't you change it, then? If you don't like what your horoscope portends, you can reconfigure it. Isn't that right? Reapportion the houses, or adjust the presumed time of birth, and . . . make it say something else. Something better. Whatever you like.

RHETICUS. I've tried that. Tried all those things. It always comes out the same.

COPERNICUS. I'm sorry, Professor. I cannot help you with your horoscope.

RHETICUS. And you call yourself a mathematician?

COPERNICUS. Exactly. I am a mathematician. Not a fortune teller.

RHETICUS. The fates of empires depend on the positions of the planets.

COPERNICUS. No, Professor. The fates of empires depend on the positions of armies on battlefields. The stars do not enter into human affairs.

RHETICUS. Tell that to your pope! Didn't he bring his favorite astrologer to Rome?

COPERNICUS. Doesn't your Luther denounce the whole practice?

RHETICUS. I told you, he knows nothing about mathematics.

COPERNICUS. Is that what you really came here for? Some new trick for casting your horoscope?

RHETICUS. Not just mine! Yours. Everybody's! Wars. Floods. Plagues. All the global predictions for the coming year. For years to come! That's what I saw as the fruit of your labors. The long march of history. The rise of Luther. The fall of Islam. The Second Coming of Our Lord and Savior Jesus Christ!

COPERNICUS. I give you the true order of the planets. The workings of the whole heavenly machinery, with every one of its former kinks hammered out. But all of that is useless to you, unless it provides excuses for every petty human failing.

RHETICUS. You think you can just twirl the Earth through the heavens like some... Like... Like... Oh, Dear Lord. Wait a minute. If the Earth moved... then... If the Earth moved through the heavens...

COPERNICUS. It does move.

RHETICUS. If the Earth moved among the planets, then it would approach them and recede from them, and that might... It would! Yes! The motion would magnify the effect of every planetary influence.

COPERNICUS. No.

RHETICUS. That would have to happen, as a natural consequence. An enhancement of the influence that each planet exerted on the individual...

COPERNICUS. The one thing has nothing to do with the other.

RHETICUS. How can you be sure? Have you checked for those effects?

COPERNICUS. No.

RHETICUS. Not even in your own chart?

COPERNICUS. No.

RHETICUS. But that would be so easy to do. To compare, say, Mars at opposition with Mars at solar conjunction, and then to...

COPERNICUS. No!

RHETICUS. This is better than I'd hoped. Better than I ever dreamed! Think what it means!

COPERNICUS. If you want to know the future, you should go slaughter a goat and examine its entrails. And leave the planets out of your predictions.

COPERNICUS *turns to leave, in the same direction as* ANNA *left.*

RHETICUS. (*restraining him*) I think there's really something to it. Let's say, just for the moment, just for argument's sake, that the Earth . . . turns. How fast would it . . . ? It has to spin around very fast, right? For the turning to cause day and night?

COPERNICUS. It is a rapid motion, yes.

RHETICUS. How rapid?

COPERNICUS. You do the math.

RHETICUS. All right. The circumference of the Earth is . . . What? Twenty thousand miles?

COPERNICUS. Twenty-four.

RHETICUS. Twenty-four thousand, right. And it would have to make a full rotation every . . . twenty-four hours.

COPERNICUS. Not a very difficult calculation, is it?

RHETICUS. God in Heaven! A thousand miles an hour?

COPERNICUS. That is what it must be.

RHETICUS. But that can't be. We would feel that.

COPERNICUS. No. We don't feel it.

RHETICUS. We don't feel it because we don't really turn.

COPERNICUS. We don't feel it because we move along with it. Like riding a horse.

RHETICUS. When I ride a horse, I feel it.

COPERNICUS. On a ship, then. Sailing on a calm sea. You move along, but you don't have any sense that you're moving.

RHETICUS. Yes, I do. I see the shore receding. I feel the breeze in my face.

COPERNICUS. Go inside the cabin, then.

RHETICUS. It won't work. It's too . . . It's . . . If the Earth turned as fast as you claim, there would be a gale, like the wind from God, howling and blowing against us all the time.

COPERNICUS. No, there's no wind.

RHETICUS. That's what I'm saying.

COPERNICUS. There's no wind because the air turns along with the Earth.

RHETICUS. The air? Turns?

COPERNICUS. It's all of a piece, yes. They turn together, as one. The Earth and the air. And the water, of course.

RHETICUS. We could not be moving that fast and not feel anything. It's impossible.

COPERNICUS. It's turning!

COPERNICUS grabs RHETICUS by the shoulders, shakes him.

All the time, it's turning.

COPERNICUS turns RHETICUS all the way around, in imitation of the Earth's rotation.

And that turning is what makes the Sun appear to rise.

COPERNICUS turns RHETICUS around, to face away from him.

And set.

> COPERNICUS *turns* RHETICUS *to face him again.*

And rise again on the following day.

> COPERNICUS *lets go of* RHETICUS, *turns away from him.*

Beat.

RHETICUS. No. It can't be. It's impossible.

COPERNICUS. You think I don't know it sounds crazy? Do you have any idea how long it took me to accept it myself? To go against the judgments of centuries, to claim something so . . . so totally at odds with common experience?

RHETICUS. Tell me about the other motion, around the Sun.

COPERNICUS. It's the same thing. You don't feel it. It's part of you, like breathing.

RHETICUS. No, I mean, is it . . . just as fast?

COPERNICUS. Oh.

RHETICUS. Is it?

COPERNICUS. No.

RHETICUS. Good.

COPERNICUS. It's faster.

RHETICUS. Damn!

Beat.

RHETICUS. *(continued)* How fast does it go?

COPERNICUS. Around the Sun?

RHETICUS. Around the Sun, yes.

COPERNICUS. I don't know.

RHETICUS. Oh, come on. Tell me.

COPERNICUS. I really don't know. No one knows the actual distance that the Earth would have to go to get all the way around, but it must be in the millions ... It must be many millions of miles. Which means we go around the Sun at least ... at *least* ten times faster than we spin.

RHETICUS. So, ten thousand miles per ...

COPERNICUS. Maybe a hundred times faster.

RHETICUS. A hundred times a thousand miles?

COPERNICUS. Maybe.

RHETICUS. That's where it all falls apart.

COPERNICUS. Why?

RHETICUS. *Why?!*

COPERNICUS. Why does it make more sense for the Sun to go around the Earth? The Sun stands still as a light for all Creation, unmoved, at the center of the universe. The way a king or an emperor rules from his throne. He doesn't hurry himself about, from city to city.

RHETICUS. But you can't put the Sun at the center. It doesn't belong there.

COPERNICUS. The Sun, in his rightful place, at the center, governs the speeds of all the planets. He tells each one how fast to run. And they obey. They take their speeds from his command. That's the reason Mercury travels the fastest.

RHETICUS. No, the Moon. The Moon travels the fastest.

COPERNICUS. Forget the Moon. The Moon is the one thing that really does circle the Earth. Mercury goes fastest around the Sun because it's the closest. And Saturn the slowest, because it's the farthest away. The planets draw some kind of motive force from the Sun's light.

RHETICUS. A force?

COPERNICUS. Yes.

RHETICUS. What kind of force?

COPERNICUS. I don't know. I am still in the dark on that matter. But it's there. And that's why all their motions are interrelated, linked together by a golden chain, and you could not alter a single one, even so much as a fraction of an inch, without upsetting all the rest.

RHETICUS. The way you talk. It's as though you know God's plan.

COPERNICUS. Why else would you study mathematics? If not to discover *that*?

Beat.

RHETICUS. And the stars?

COPERNICUS. They hold still, the same as the Sun. The sphere of the stars can't spin around the Earth every day. It's too big.

RHETICUS. I'm trying to see it your way. Really, I am. But, if the Earth moves around the Sun . . . Shouldn't we see some change in the stars? Wouldn't some of them look . . . I don't know . . . closer together sometimes, or farther apart? There should be a change, from spring to fall, that people who paid attention would notice.

COPERNICUS. You would think that would happen.

RHETICUS. I don't know what to think.

COPERNICUS. But no. You don't see any seasonal difference. Because the stars are so much farther away than anyone has imagined. The scale of the universe is all but inconceivable. The distance to the stars is so tremendous that it dwarfs the distance between the Earth and the Sun. Compared to the distance from

Saturn to the stars? Why, the distance from the Earth to the Sun is . . . negligible.

RHETICUS. Negligible?

COPERNICUS. It shrinks to just a point, really.

RHETICUS. You're making this up. It's your own fantasy. The stars get in your way? You just wave them off to some other place.

COPERNICUS. Don't impose any puny, human limits on Creation. As though the whole cosmos were just a crystal ball for your own little personal affairs.

RHETICUS. In the name of the Creator, then: What is the use of all that empty space between Saturn and the stars?

COPERNICUS. The *use*?

RHETICUS. Yes.

COPERNICUS. What is the use of grandeur? Of splendor? Of glory? Thus vast, I tell you, is the divine handiwork of the one Almighty God!

Blackout. End of Act I.

A native of Torun, Copernicus lived thirty years in Frauenburg, "the city of Our Lady," in the shadow of its medieval cathedral. Frauenburg, the seat of the Varmia diocese, is the setting for the play.

ACT II

Scene i. Outside Copernicus's house

COPERNICUS. I wish you would stay a while longer.

GIESE. It's better this way. Once Johann gets hold of an idea . . . Out of sight, out of mind.

COPERNICUS. I promise you, Tiedemann, your canonry is safe. The Chapter will never act to remove you.

GIESE. Still, I'd best be going. Remember what I said. Should it become necessary, the position is a good one, with a fine family. She'd be well treated there.

COPERNICUS. I can't just dismiss her, like an ordinary servant.

GIESE. You'll find the right words to do the right thing.

COPERNICUS. If I were sure it was the right thing.

GIESE. There's something else I want to say, Nicholas. About your visitor.

COPERNICUS. You haven't told anyone who he is?

GIESE. I can hardly believe his identity myself. The way he materialized on your doorstep.

COPERNICUS. He's almost ready to leave us.

GIESE. A professor of mathematics, praise God.

COPERNICUS. A professor from Wittenberg.

GIESE. The gifts he brought you. Texts by the ancient sages. And those letters from your former correspondents, with all of them still eager to hear the exposition of your theory.

COPERNICUS. No. Not really. They don't even understand what I...

GIESE. And as though all that weren't enough, he hand-delivered the very information you lacked for your calculations.

COPERNICUS. That was a welcome surprise, yes.

GIESE. Well?

COPERNICUS. Well, what?

GIESE. What more do you want? A lightning bolt? An angel to knock you on the head?

COPERNICUS gestures his bewilderment.

GIESE. *(continued)* Open your eyes, Nicholas. His arrival is a sign.

COPERNICUS. A sign? A sign of what?

GIESE. Ah, Nicholas. Of all the literary and artistic pursuits, yours is the one above all others to be embraced and pursued, because it concerns the most beautiful and worthy objects: the majestic circular movements of the world and the course of the stars.

COPERNICUS. But Tiedemann...

GIESE. Think of it, Nicholas. The two of you, together. Catholic and Lutheran. Mentor and pupil. The wisdom of age and the energy of youth, united in fruitful discourse by the force of a powerful vision. And the world enlightened by your collaboration.

COPERNICUS. You can't be serious, Tiedemann. You think God has sent me a Lutheran to convince the world the Earth turns?

GIESE. You were called once, to craft your theory. And now,

Nicholas, with our Mother Church shaken by discord, Heaven calls you again.

Blackout.

Scene ii. Inside Copernicus's house

RHETICUS *sits reading the piles of pages that* COPERNICUS *and* ANNA *took from the trunk.*

ANNA *enters.*

ANNA. I see you're feeling better.
RHETICUS. Why, yes. Thank you.
ANNA. Then I trust you'll be leaving us soon.
RHETICUS. Well, in fact I'm hoping to stay. And help Canon Copernicus with his work.
ANNA. You can't do that. Don't you understand? We just want to be left in peace.
RHETICUS. I appreciate your concern. But, still, I think . . .

COPERNICUS *enters.*

RHETICUS. *(continued)* Oh, there you are, Canon, Sir.
ANNA. I was just helping him prepare for his journey. I'm going out now to make my calls. *(to* RHETICUS*)* Goodbye, then.
COPERNICUS. Goodbye, my dear.

ANNA *exits.*

RHETICUS. I hope you don't mind: I could not help seizing the opportunity to read a bit deeper into your manuscript.

COPERNICUS. Oh?

RHETICUS. It is a phenomenal piece of scholarship, Sir, if I may say so.

COPERNICUS. Thank you.

RHETICUS. Even if one does not concur with your reasoning... Even if one has trouble conceding motion to the Earth, still there is something so... I wonder, Canon, Sir, if you can tell me exactly what inspired you to swap places between the Earth and the Sun?

COPERNICUS. What steps?

RHETICUS. Yes. Yes exactly.

COPERNICUS. I don't think so. No. The circles just seemed to allow the possibility. And to fit together better that way. As though all the pieces had found their proper places. Do you see?

RHETICUS. Ah. Well, no one could gainsay your calculations. Nor the effort, the years of observations you provide alongside your calculations. No, Canon, Sir. On the whole, the work seems more than the intellect or the labors of a single individual could accomplish in a lifetime. And yet somehow you accomplished it. Have you really had no help at all?

COPERNICUS. Only some... *amateur* assistance.

Beat.

RHETICUS. You know those books I brought you?

COPERNICUS. Do you want them back? I said you could keep them.

RHETICUS. Did I mention that I am acquainted with that printer? I met him in Nuremberg. He is excellent.

COPERNICUS. Indeed so.

RHETICUS. The best printer of scientific texts in all of Germany. Perhaps in all of Europe.

COPERNICUS. I don't doubt it.

RHETICUS. He is just the man to publish your book.

COPERNICUS. My book?

RHETICUS. No one else comes to mind who has the capacity and experience to reproduce charts and diagrams like yours. Why, the numerical tables alone pose a considerable technical challenge for a printer.

COPERNICUS. But I told you I'd decided against publication.

RHETICUS. Oh, it needs more work, I agree. Some reorganization of the various sections. And a judicious rewording, of course, of the more jarring notions. But you could publish this. What I mean is, I think you should publish it. In fact, you must.

COPERNICUS. That's easy for you to say. You would not face the scorn that I have to fear.

RHETICUS. The mathematicians may raise their hackles. True. But when they grasp the practical advantage of your algorithms . . .

COPERNICUS. Not just mathematicians. Church men will oppose me. Already I have had Joshua raised against me so many times that I feel myself like one of his enemies among the Amorites.

RHETICUS. After you publish your book, if someone disagrees with you . . .

COPERNICUS. *"If"* someone disagrees? *If?!*

RHETICUS. If someone disagrees with you, let him publish a counterargument. Then you come back to refute his counterargument. And you go on like that. Back and forth. That's how learned men make good use of the God-given printing press.

COPERNICUS. No. A fog of absurdity hangs over my theory.

RHETICUS. It would be wise to avoid discussing what you've done to the planetary order. And focus instead on the value of the calculations. The calculations, on their own, would enable future astronomers to . . .

COPERNICUS. But Professor, the circuit of the Earth around the Sun is what sets my work apart. It restores all the planets to perfect, circular, and uniform motion. Its great strength lies in the way it unifies the whole system of the heavens. All the other theories divide the spheres. They apply a different principle to each planet, which is why they fail utterly to paint a true picture of the heavenly form and symmetry. Why, you might as well try painting a person by copying a head from here, hands from there, and feet from somewhere else. In the end, though each part were handsome in itself, together they would make a monster instead of a man.

Beat.

RHETICUS. To present your theory as the true workings of the heavens, is the same as claiming these ideas have been divinely revealed to you. Are you prepared to say that?

COPERNICUS. I don't want to say any of it. I choose to follow the example of Pythagoras. He kept his secret numbers a secret. He never divulged them to anyone, except his kinsmen and friends. And then only by word of mouth. Never in writing.

RHETICUS. He was afraid someone would steal his idea.

COPERNICUS. That's not what he was afraid of. I know how he felt. He wanted to protect his most beautiful thoughts from ridicule.

RHETICUS. Secrecy has no place in our philosophy. This work must be shared.

COPERNICUS. I started out with the intent to share it, but then I...

RHETICUS. Let me help you finish what you started.

COPERNICUS. Help from you? Even though you reject the very heart of my theory?

RHETICUS. Whether or not I agree with you, you cannot keep this to yourself any longer. It's too important.

COPERNICUS. Even if I were to accept your help, think of the risk. I would have to hide you. Lie about you. We'd be in constant danger.

RHETICUS. For the chance to build a boundless kingdom in astronomy, what risk would be too great?

Blackout.

SCENE iii. BISHOP'S PALACE

BISHOP. Has he recovered, then?

COPERNICUS. He has, yes.

BISHOP. What have you learned about him?

COPERNICUS. It's rather something he taught me, about myself, Your Reverence.

BISHOP. Oh, yes?

COPERNICUS. He has awakened in me a wish to resume my own work in mathematics.

BISHOP. But who is he?

COPERNICUS. The explication of my theory. I'm thinking I should publish it.

BISHOP. I would advise against that, Nicholas. I don't think it's a wise decision. Your idea is . . . Well. I'm sure it's very mathematical. But, at the same time, it oversteps the bounds of mathematics. As I see it, I will go so far as to say it shakes the very foundation of Church doctrine.

COPERNICUS. Oh, no, Your Reverence.

BISHOP. What about Joshua?

COPERNICUS. Ah.

BISHOP. Well? Answer me. What about him? How will you defend yourself against the charge that your ideas conflict with the Biblical account of Joshua?

COPERNICUS. I think it's best to say nothing.

BISHOP. You refuse to answer me?

COPERNICUS. Oh, not you, Your Reverence. I don't answer the charge. I would rather avoid any mention of Joshua. And limit my comments to mathematics alone.

BISHOP. That's no answer.

COPERNICUS. I'm afraid, Your Reverence. Afraid there may be . . . babblers, who claim to be judges of astronomy, although completely ignorant of the subject. And such men are not above twisting some passage of Scripture to their purpose, to censure me.

BISHOP. I am not trying to censure you.

COPERNICUS. Oh, I know Your Reverence is not one of those.

BISHOP. I have overlooked all sorts of infractions lately, as I need not remind you!

COPERNICUS. Your Reverence is most kind.

BISHOP. But you will most certainly have to deal with Joshua. And many other passages of Holy Writ. The Psalms also teach us that the Earth does not move.

COPERNICUS. As I read those passages, I hear the godly Psalmist declare that he is made glad through the work of the Lord. That he rejoices in the works of His hands. Only that.

BISHOP. Are we reading the same Bible, Nicholas? Psalm 104 says the Lord God laid the foundation of the Earth, that it not be moved forever. Not be moved. Forever.

COPERNICUS. It's a matter of interpretation.

BISHOP. What's to interpret? It's stated there in plain language. It couldn't be more clear. Not to be moved forever. It doesn't say it should spin like a top.

COPERNICUS. To me, it says that God, the source of all goodness, created an abiding home for mankind on this Earth. And that foundation will hold firm forever.

BISHOP. That still doesn't answer Joshua.

Beat.

COPERNICUS. As strange as this may sound to someone who is not a mathematician, Your Reverence, my theory offers certain advantages for the improvement of the calendar.

BISHOP. The Ecclesiastical calendar?

COPERNICUS. Easter, for example. To calculate the correct date of Easter each year.

BISHOP. You could make a contribution of that significance?

COPERNICUS. I don't mean to boast.

BISHOP. Why didn't you say so before? Why did you never mention the calendar until now?

COPERNICUS. I lacked the confidence to expose my theory. To the scrutiny of others.

BISHOP. I had no idea there was so much to it.

COPERNICUS. Then . . .

BISHOP. All these years, I thought you were just . . .

COPERNICUS. Do I have Your Reverend Lordship's blessing to continue the work?

BISHOP. I suppose so. If what you say is true, then I suppose you *should* take it up again.

COPERNICUS. Thank you, Your Reverence.

BISHOP. How does that work, then? How does what you do relate to the date of Easter?

COPERNICUS. It concerns correcting the exact duration of the tropical year, from equinoctial and solstitial observations.

BISHOP. Never mind.

COPERNICUS. And now I suppose I'll need to find a press. A printer. I hear there is an excellent one in Germany.

BISHOP. Germany! Have we no printers here in Poland?

COPERNICUS. None, I think, that could take on a work of this nature.

BISHOP. Such a book could bring very positive attention to Varmia. Not just Varmia. To Poland. To . . . It should definitely be printed here.

COPERNICUS. It's such a lengthy work . . .

BISHOP. What do you know of printers in Germany?

COPERNICUS. For this kind of text, with the large number of diagrams and numerical tables required . . .

BISHOP. Aren't you getting ahead of yourself? Don't you think you should write this book before you worry about where to have it printed?

COPERNICUS. Yes, Your Reverence. There is much to be done.

BISHOP. So it's a big book, is it?

COPERNICUS. I've written several hundred pages, over the years.

BISHOP. As much as that?! My, my. And still not come to the end?

COPERNICUS. It's . . . complex.

BISHOP. A scholarly book like this will surely attract the duke's attention. He'll see what talents we have here in Varmia. Why, the king himself might recognize the diocese for such a . . . How long do you think it will take you?

COPERNICUS. That is difficult to say. I'll need help to complete the project.

BISHOP. You may have my secretary. I'll put him at your disposal.

COPERNICUS. That's not the kind of help that I . . .

BISHOP. Ah! You mean the mathematics.

COPERNICUS. Yes, Your Reverence.

BISHOP. I'm afraid I can't help you with that.

COPERNICUS. I suspect, Your Reverence, that the unfortunate traveler who recently fell ill at my door, might be . . .

BISHOP. Back to him again. You haven't even told me his name.

COPERNICUS. It's Rheticus, Your Reverence.

BISHOP. Rheticus?

COPERNICUS. Professor Rheticus, yes.

BISHOP. And where does he teach? At Krakow?

COPERNICUS. If he were amenable, and if Your Reverence would allow, I might ask him to stay on a while, as a collaborator.

BISHOP. Since when have you sought my approval of your house guests, Nicholas? Or your housekeeper?

COPERNICUS. This is an altogether different situation.

BISHOP. You flout my wishes and then ask for more concessions? You can't have them both, Nicholas. The harlot. Or the helper. One or the other.

Blackout.

SCENE iv. INSIDE COPERNICUS'S HOUSE

All evidence of RHETICUS *and his belongings, as well as the trunk and manuscript, have been removed (taken to the Tower).*

ANNA. I don't like it, Mikoj. You know I would never stand in the way of your work, but just his presence here . . . It worries me.

COPERNICUS. Anna. I . . .

ANNA. Haven't we enough trouble?

COPERNICUS. Indeed, we do. That's what I want to tell you, Anna.

ANNA. What?

COPERNICUS. Perhaps it would be wise for you to be away for a little while.

ANNA. Away? Me?

COPERNICUS. Just a little while. We'll hardly have time to miss one another.

ANNA. I'm not going away, Mikoj. I told you I'd never leave you, and I meant it.

COPERNICUS. Just until the bishop calms down.

ANNA. You said you wouldn't let him send me away.

COPERNICUS. Anna, I feel I must bring my work to completion. You know, better than anyone, how hard we worked to get as far as we did. But now I need help from someone else. And I can have that help only if you go away. Just until the work is done. And then we can be together again.

ANNA. Oh, Mikoj. I don't know. What would I do? Where would I go?

COPERNICUS. Tiedemann Giese has offered a place for you with one of his . . .

ANNA. A place for me?

COPERNICUS. Yes, one of his parishioners needs someone to . . .

ANNA. You have a place for me? Already?

COPERNICUS. It's just that he wanted to help us, to make it easier to . . .

ANNA. So! Then it's all arranged? Everything settled?

COPERNICUS. I wouldn't add that to your burden. Of course I looked into another situation for you. A temporary situation.

ANNA. I don't want another situation. This is my home. Whereas that professor . . . If the bishop knew the truth about him . . .

COPERNICUS. You wouldn't tell anyone, would you? That he's a . . . ?

ANNA. Oh, Mikoj! How could you?
You know I would never do anything to hurt you.

COPERNICUS. Forgive me.

They embrace, comfort each other, until ANNA *pulls back.*

ANNA. Come with me.

COPERNICUS. What?

ANNA. Come away with me. Let's both go. Why should you grovel to him any longer?

COPERNICUS. You think I could simply walk away from here?

ANNA. Run away. Come and be with me where no one will care that we're together. Leave all these nasty, nosy old men.

COPERNICUS. Leave the Church?

ANNA. We'll make a new life. Our own life. Think of it, Mikoj. We could have a hospital. We'll get by. You'll see.

COPERNICUS. I'm too old to change, Anna.

ANNA. You mean you won't go with me?

COPERNICUS. I can't.

ANNA. You can't?

COPERNICUS. I'm sorry. Anna . . .

ANNA. What do you mean, you can't? You turned the whole universe inside out. You showed every planet which way to go. Tell me, Mikoj, are you still that man?

Blackout.

SCENE V. TOWER ROOM

The Tower has been tidied, with the table and chair used as a desk, and several hundred manuscript pages stacked here and there.

RHETICUS. It's better now, the way you've changed it. But I still say you make the case too quickly.

COPERNICUS. I can't pretend not to know what I know.

RHETICUS. Yes, but must you begin by saying the Earth moves and the Sun stands still?

COPERNICUS. I feel I must, yes.

RHETICUS. Why not wait a few pages? Rather than rush right into the Sun, start with some other . . . Where is that section about . . . Here! This part, where you explain how you approached the cycle of the eighth sphere.

COPERNICUS. It makes no sense to start there.

RHETICUS. You insist on pushing the Sun into the center of the universe on page one?

COPERNICUS. That is the basis of the whole structure.

RHETICUS. I still say you build up to it, the way I tried to show you.

COPERNICUS. But I don't . . .

RHETICUS. You cannot pluck the Sun, the very lantern of the universe, from the eternal, perfect heavens, and shove it into this hellhole of a pit where the Earth now wallows.

COPERNICUS. But I explain how it compels the planets . . .

RHETICUS. The philosophers will insist on keeping the Earth at the center because . . . Because of its Earthiness. Because of all the change, and death, and decay in this realm. If you want to put the Sun in the midst of all that, you had better do it slowly.

Beat.

COPERNICUS. Why don't you say what you really mean? I haven't proven it.

RHETICUS. I didn't say that.

COPERNICUS. But that's what you mean. If the proofs were stronger, you wouldn't be trying so hard to make it sound palatable.

RHETICUS. I want them to hear you out, to see what you've done. I'm begging you: Invite them into this new world. Don't foist it on them.

COPERNICUS. Maybe it isn't ready after all. Maybe this was all a big mistake.

RHETICUS. No, no. Don't lose heart.

COPERNICUS. I don't know what made me think I could . . .

RHETICUS. You've got to leave a few stones unturned. Something for others who come after you to do. You've given us so much to build on. Your work is . . . It's like that cathedral out there. Do you think anyone who laid the stones for the foundation was still around when the cross went up on top? Trust me, Father. A hundred years from now, astronomers will still be reading your book.

COPERNICUS. And you, Joachim?

RHETICUS. I shall read it a hundred times.

COPERNICUS. What will you do after we finish here?

RHETICUS. After? Why, take the manuscript to Nuremberg, of course. I'll watch over the printer, to keep him honest. I'll proof-read every page, I'll . . .

COPERNICUS. *After* that.

RHETICUS. I don't have to worry about anything after that.

COPERNICUS. You'll go back to Wittenberg? To your teaching?

RHETICUS. No, Father. By that time . . . By then I'll be . . .

COPERNICUS. What?

RHETICUS. There is no "after" for me, after that. Don't you remember? By this time next year, when Jupiter and Saturn enter their Great Conjunction, my time will be . . .

COPERNICUS. You still fear that?

RHETICUS. Nothing in your theory gives me a way out.

COPERNICUS. You can't just resign from life. Acquiesce to some fate you think awaits you.

RHETICUS. In these weeks we've spent together, I accomplished my mission. Not many older men can say as much. I found you. I rescued your work from oblivion. And once I see it published, I'll have done all I can do. It will no longer matter what happens to me.

COPERNICUS. You don't know what will happen.

RHETICUS. But I do.

COPERNICUS. You could live a hundred years. You have no idea what the future holds.

RHETICUS. You did everything you could for me. The time with you has been the most . . .

COPERNICUS. Wait and see what happens to your career when Schöner and the rest of them read my acknowledgments to you.

RHETICUS. To me?

COPERNICUS. Of course to you.

RHETICUS. Oh, no. You mustn't disclose my role in this.

COPERNICUS. You think I wouldn't thank you, publicly, for all you did to . . . ?

RHETICUS. My name must not appear in your book. It would taint the whole thing.

COPERNICUS. I don't care. I owe you so much.

RHETICUS. No. You have others you must thank. I have a plan, for a dedication that you will write. To the real power.

COPERNICUS. You mean Duke Albert?

RHETICUS. No.

COPERNICUS. The king?

RHETICUS. No, no one from the government. The dedication must acknowledge higher powers. Someone in the Church.

COPERNICUS. You mean the bishop?

RHETICUS. No!

COPERNICUS. Tiedemann Giese?

RHETICUS. Not nearly powerful enough.

COPERNICUS. Who then, the pope?

RHETICUS. Yes!

COPERNICUS. I was joking, Joachim.

RHETICUS. I am perfectly serious.

Beat.

COPERNICUS. His Holiness?

RHETICUS. He's really the only one. To protect you. From those backbiters who will bend chapter and verse to evil purposes, and try to condemn your theory. Even though, as we both know, there is nothing irreverent in your book, there is always the danger that someone will twist a passage of Scripture . . .

COPERNICUS. But . . . His Holiness.

RHETICUS. The mere mention of his name will lend the book the air of papal authority. It might even give people the impression that he had commissioned you to write it.

COPERNICUS. He would never do that.

RHETICUS. Still, it might appear that he had.

COPERNICUS. What could he possibly say about astronomy?

RHETICUS. He need not say anything. You simply dedicate the book to him.

COPERNICUS. I could not do even that much without his express permission.

RHETICUS. Then we must get his permission.

COPERNICUS. He has the troubles of the world on his shoulders. Why, he's gone and excommunicated the king of England. He is consumed with crushing the Lutheran heresy! I'm sorry, Joachim. Forgive me, but . . .

RHETICUS. Trust me, Father. If you dedicate your studies to him, then you prove to everyone, that you do not run away from judgment, even by the highest authority.

COPERNICUS. The audacity of it.

RHETICUS. Your bishop must have representatives in Rome. Ambassadors to the Vatican? Someone who can gain an audience with him?

Beat.

COPERNICUS. (*starts to laugh*)

RHETICUS. What?

COPERNICUS. (*laughs harder*)

RHETICUS. (*laughing, too*) What?

COPERNICUS. I'm just trying to picture the bishop's face when I ask him to . . .

RHETICUS. Ha! (*both of them laughing*)

Blackout.

SCENE vi. BISHOP'S PALACE

BISHOP *paces with excitement, planning the liaison with Rome.*

BISHOP. Paul Pontifex Maximus the third.

COPERNICUS. Yes, Your Reverence.

BISHOP. His Holiness! The imprimatur of Rome bestowed on a text from Varmia.

COPERNICUS. I'm happy the idea finds such favor with Your Reverence.

BISHOP. It's just as I told you, Nicholas. Now you see how getting rid of that hussy has freed your mind for the serious work God intended you to do.

COPERNICUS. I stand ever indebted for the patience and beneficence Your Lordship has granted.

BISHOP. You know what I'm thinking, Nicholas? I'm thinking you'll need to leaven your mathematical jargon with a little poetry. Your magnum opus will require that literary touch. Yes. And I shall compose an elegant epigram for you, to serve as a poetic invocation.

COPERNICUS. Your Reverence is too kind.

BISHOP. I must set my mind to it. It's been quite a while since I tried any versifying.

COPERNICUS. Take as long as necessary, Your Reverence. We are several weeks away from completion.

BISHOP. Still not finished? Even with all the help you've had from your Professor . . . Hereticus is it?

COPERNICUS. Rheticus, Your Reverence.

BISHOP. That's right. Well, I suppose he is to be congratulated for bringing your work to this stage.

COPERNICUS. Indeed, Your Reverence. I could not have done this without his help. I shall miss him when he returns to Wittenberg.

BISHOP. Wittenberg?!

COPERNICUS. Remember, I did not invite him here, Your Reverence.

BISHOP. A Lutheran?!

COPERNICUS. As God is my witness, I had no inkling he was coming. He materialized on my doorstep like a . . . like . . .

BISHOP. How you insult me! Abuse me!

COPERNICUS. It was all such a coincidence that it seemed it could not be merely coincidence . . . As though there were something . . . Providential in his arrival.

BISHOP. Silence! Heaven does not dispatch Lutherans to do our bidding in Varmia! Damn it all, Nicholas! You know the law. Now the law must be enforced.

Blackout.

SCENE vii. TOWER ROOM

COPERNICUS *has just returned to inform* RHETICUS *of the danger.*

RHETICUS. Now?

COPERNICUS. We've run out of time.

RHETICUS. But we're almost finished.

COPERNICUS. He's going to have you arrested.

RHETICUS. Just a few more days.

COPERNICUS. I can't protect you from the King's soldiers. You must leave, Joachim.

RHETICUS. I can't abandon you now.

COPERNICUS. I won't have you risk your life for this.

RHETICUS. We're so close. Another few days and we . . .

COPERNICUS. No. Not another word. Off with you now.

RHETICUS. If that's what you want.
COPERNICUS. I'm so afraid for you, Joachim.
RHETICUS. All right then, I'll go. For your sake.

RHETICUS begins gathering the piles of manuscript pages.

COPERNICUS. What are you doing?
RHETICUS. I'll take the book to Nuremberg. Just as we . . .
COPERNICUS. No. You can't . . .
RHETICUS. To the printer.
COPERNICUS. No.
RHETICUS. I will keep my promise.

RHETICUS continues packing the manuscript.

COPERNICUS tries to stop him, grabs the pages from him.

COPERNICUS. Stop!
RHETICUS. What's the matter with you?
COPERNICUS. It's not ready.
RHETICUS. I know, but we can continue to work on the rest in correspondence with each other.
COPERNICUS. You cannot take my manuscript!
RHETICUS. I promise I will guard it with my life.
COPERNICUS. No. No. I never meant for you to take it away.
RHETICUS. We've been working toward this moment all along.
COPERNICUS. A copy. I meant for you to take a copy with you. Not my manuscript.
RHETICUS. But I've copied only the first few chapters. Not enough to . . .

COPERNICUS. I can't. I need to keep it here. It's been with me my whole life. It *is* my life.

> RHETICUS *stops struggling.*

> COPERNICUS *sinks into a chair, spent, panting.*

COPERNICUS. (*continued*) I cannot part with it.
RHETICUS. And I cannot let it languish here, unseen. If it has been your life, then let it be your legacy. Let it come to new life in the minds of other men.
Beat.
You've trusted me this far. Trust me now.

> COPERNICUS *recovers himself, stuffs pages into* RHETICUS's *satchel.*

> RHETICUS *also resumes packing the pages.*

COPERNICUS. Make sure you have it all.
RHETICUS. I will not fail you.
COPERNICUS. Hurry now.
RHETICUS. I'll see it through.
COPERNICUS. Yes. Now go.
RHETICUS. Goodbye, then.
COPERNICUS. We're not likely to see one another again.

> COPERNICUS *and* RHETICUS *embrace.*

RHETICUS. God be with you, my teacher. My father.

> RHETICUS *exits.*

COPERNICUS. And with you. May God protect you.

Blackout.

SCENE viii. INSIDE COPERNICUS'S HOUSE

Months later, COPERNICUS *lies in bed, comatose.* GIESE *kneels beside him, praying.*

A knock at the door.

GIESE *goes to answer.*

ANNA *enters.*

ANNA. Where is he?

ANNA *rushes past* GIESE *to the bedside.*

ANNA. Oh, Mikoj! It's me, my dearest. I'm here with you now. It's all right. The bishop tried to keep me away, but I'm here now. And I'll stay with you. I'll be here every minute. Don't worry. I'm here.
GIESE. Alas, he doesn't hear you.
ANNA. Shh. Look! He's trying to speak.
GIESE. He hasn't said a word in weeks. Nothing.
ANNA. But his eyes are open. His lips are moving. Look.
GIESE. The Duke sent his personal physician. He said it's just a . . . a reflex.
ANNA. We don't know that. He may hear everything we're saying. Can you hear me, Mikoj? You don't have to talk if you

don't want to. If it's too hard for you, you just rest. I'm not leaving you now.

GIESE. I have administered the last rites. God will call him when it's time.

ANNA. That book was the death of him.

GIESE. No, my child. It was good for him to finish it. To let it go out into the world. And once it did, then his work on this Earth was completed.

RHETICUS enters.

RHETICUS. Am I too late?

ANNA. What are you doing here?

RHETICUS. I came as soon as I could. I wanted him to know that I . . . I wanted to show it to him myself.

GIESE. What do you have there? Is that . . . ?

RHETICUS. His book, yes. I brought it.

GIESE. Let me see it. Oh, my. I can hardly believe my eyes.

ANNA. That's it? That?

GIESE. I never thought I'd see the day.

ANNA. That can't be it. Just a pile of paper?

GIESE. "*On the Revolutions of the Heavenly Spheres.* By Nicolaus Copernicus."

ANNA. It's not at all what I imagined. How shabby it looks. He deserved better from you.

GIESE. This is the way books come from the printer. Just the pages, like this. But I shall have it bound for him. Something very grand. Red leather, with his name stamped in gold letters.

Act II, scene viii · 81

RHETICUS. I wanted to do that myself, but I was rushing here . . . Hoping . . .

GIESE. All the times I urged him to do this. And how he fought against me. The stubborn old mule. I wish he could have seen this.

ANNA. Let's show it to him now. Give me some of the pages.

ANNA *takes the pages to* COPERNICUS; *stays by him.*

GIESE. This was a brilliant stroke, the dedication to His Holiness.

ANNA. Wait till you see what's here, Mikoj.

GIESE. Wait a minute . . . Where are Johann's verses?

RHETICUS. They were doggerel. Not fitting at all for his masterpiece.

GIESE. And what is this "Note to the Reader"? I don't remember his writing anything like this.

ANNA. It's your book, Mikoj. Your very own book.

GIESE. Who did this?! Who wrote these . . . lies?! Right here at the very beginning.

RHETICUS. I can explain.

GIESE. (*reading aloud*) "These hypotheses need not be true, nor even probable"? (*to* RHETICUS) What have you done? How could you undermine him in this egregious fashion?

RHETICUS. I didn't write that.

GIESE. Who then? Who else had access to the press?

ANNA. All your work, all those years, and here it is, at last.

GIESE. (*reading aloud*) "Let no one expect the truth from astronomy, lest he depart from this study a greater fool than when he entered it."

ANNA. Touch it, Mikoj. Isn't it wonderful?

> ANNA *tries to put the pages in* COPERNICUS'S *hands.*

GIESE. How dare you? He trusted you. And you betrayed him!

RHETICUS. No!

GIESE. The villainy! How could you say such things? Twist his meaning? Deny his vision?

RHETICUS. But I didn't. The printer put that in. He insisted on it.

GIESE. And you let him?

RHETICUS. He refused to print the book without such a disclaimer.

ANNA. Mikoj?

GIESE. Coward! Ingrate! You make me glad he cannot see what a travesty has been made of his life's work. You must go back to the printer, make him do it over, set it right.

RHETICUS. No. It's better this way. The mathematicians will welcome the book the way it is. Now they have permission to ignore what they are not prepared to accept.

GIESE. Ignore it?

ANNA. Mikoj?

RHETICUS. Let them cling to the immobile Earth. Let them take what they can use of his ideas, and leave the rest.

ANNA. Mikoj!

> *The pages drop from* COPERNICUS*'s hand and fall to the floor.*

GIESE. The Lord has taken him. God rest his soul.

RHETICUS. His good soul. Amen.

> GIESE *kneels, prays over* COPERNICUS.
>
> RHETICUS *also prays, standing, looking up.*
>
> ANNA *remains kneeling by* COPERNICUS.

GIESE. Requiem Aeternam dona eis, Domine et lux perpetua luceat eis: Requiescant in pace. Amen.

> GIESE *finishes with the sign of the cross.*

RHETICUS. Did you believe him?
GIESE. I loved listening to him talk about it.
RHETICUS. Yes, but did you . . . ?
GIESE. Every time I open his book, I will feel his spirit near me.
RHETICUS. But do you believe the things he wrote? Believe what he believed?
GIESE. No one can know the true structure of the heavens, unless it has been divinely revealed to him.
RHETICUS. Do you think it was? Revealed to him?
GIESE. I do not dismiss that possibility. Surely there is something majestic in the beauty, the unity . . . the perfection of what Nicholas conceived.
RHETICUS. Yes.
Beat.
ANNA. I remember the time we watched an eclipse together, out there in the meadow. The Moon was full, and everything so bright. You could have read this book by the moonlight. I must have drifted off to sleep. Because he had to wake me when the shadow appeared. He couldn't leave his instruments, of course,

so he made a howling sound. Like a wolf. Then everything moved slowly, as in a dream. It must have taken an hour, I think, for the shadow to creep across the Moon's face, and color it that gorgeous, coppery shade of red. The most beautiful sight I ever saw.

Beat.

GIESE. I, too, recall moments like that, with him.

RHETICUS. As do I.

GIESE. At such times, as God is my witness . . .

GIESE shakes his head, unable to complete the thought.

RHETICUS. I know. It happens to me, too . . . Sometimes, when I think of him, when I hear his voice in my head, I would swear, I can almost feel it turn.

Blackout. End of play.

Afterword

LETTERS EXCHANGED AFTER THE fact among Copernicus's intimates describe his demise. He apparently suffered a stroke in the autumn of 1542. By then, Rheticus had delivered a copy of the manuscript to the printer Johannes Petreius of Nuremburg, and was periodically sending batches of printed pages to Copernicus for review. After the stroke, Copernicus could no longer proofread and comment on them. Unable to speak, and partially paralyzed, he held on for months in the care of his fellow canons—until the final installment of his book reached him by courier in late May, and he died with the pages in his hands.

Copernicus's melodramatic end meant that he never learned the world's reaction to *On the Revolutions of the Heavenly Spheres*. Nor did he realize that an anonymous preface had been inserted in the front matter.

Copernicus's dear friend Tiedemann Giese, outraged by the new preface, tried to sue the printer, but to no avail. Rheticus crossed out the offending preface with a red crayon in the copies he gave as gifts to friends. Eventually the preface was attributed to Andreas Osiander of

Nuremberg, a Catholic priest turned Lutheran theologian, with friendly ties to Petreius the printer.

Several early readers approached *On the Revolutions* as the anonymous preface advised—as an aid to calculation that bore little relation to the real structure of the heavens. These readers ignored Copernicus's blueprint for "the movements of the world machine." Meanwhile Rheticus continued, until his death in 1574, to remind mathematicians that Copernicus's "hand advanced the machinery of this world." (I hasten to add, however, that the mechanical "World Machine" in the play is my own invention. I am certain Copernicus never saw the need to build any such device for demonstration.)

Johannes Kepler became Copernicus's next outspoken, fully convinced convert toward the end of the sixteenth century. Kepler had access to the trove of precision data on planetary positions amassed in Denmark by Tycho Brahe, and used those measurements to discern the true nature of the planets' orbits. Kepler authored several persuasive texts in which he defended Copernicus, claiming, "I lay my whole astronomy upon Copernicus's hypotheses concerning the world." He urged his cautious Italian correspondent, Galileo Galilei, to do the same, and to speak out in favor of the moving Earth.

By the time Galileo made the telescopic discoveries that emboldened him to endorse Copernicus in public, *On the Revolutions* had been in print for nearly seventy years. No ridicule or religious outcry had yet assailed it.

Possibly the Osiander preface served to diminish the theory's shock value, and the book's erudition (it was written in Latin for university-educated mathematicians) limited its readership. The Copernican model did not become a cause célèbre until Galileo began promoting it to a wide audience. In response, in 1616, the Sacred Congregation of the Index judged Copernicus's tenets "false and contrary to Holy Scripture." In 1620, the Congregation's cardinals issued corrections to be entered in the text by all owners of the book, so as to make its conclusions sound more like conjectures. Galileo's copy shows that he dutifully crossed out or edited all the offending passages.

Silenced, Galileo said nothing further about Copernicus until the election of a new pope in 1624 encouraged him to reopen the controversy. Galileo's *Dialogue Concerning the Two Chief Systems of the World*, published in 1632, precipitated his trial by the Holy Office of the Inquisition. His *Dialogue* was banned in the trial's aftermath; it remained on the *Index of Prohibited Books* for the next two centuries.

Given that all five characters in *And the Sun Stood Still* lived real lives, and that four of them were published authors, I was able to put a number of their own words in their mouths. One such borrowing is Copernicus's pronouncement at the end of Act I, "Thus vast, without any question, is the divine handiwork of the one almighty God." This statement, from book one of

On the Revolutions, was also singled out by the Sacred Congregation of the Index—for deletion. Though pious in sentiment, it boasted that Copernicus somehow knew the true dimensions of the universe.

Today, *On the Revolutions* in its many modern translations still attracts curious readers. The Latin original holds iconic value for historians of science and collectors alike. Owen Gingerich located 277 copies of the 1543 first edition in his global survey, as well as 324 copies of the second edition, which was printed in Basel in 1566. The many volumes bear scant resemblance to one another, despite their shared text, because each owner bound his copy according to personal taste and budget. A "very crisp and clean" first edition in a contemporary binding of flexible vellum sold at auction in New York in June 2008 to an undisclosed buyer for $2,210,500.

As for Copernicus himself, the archaeologists in 2005 succeeded in unearthing the skull and bones of a seventy-year-old man from the crypt beneath the Frauenburg—now Frombork—cathedral. After various tests to confirm the identity of the remains, Canon Copernicus was reburied on May 22, 2010, with the primate of Poland presiding at the Mass.

The grand vision Copernicus conceived of a heaven too vast for human imagining continues to expand all around us.

In his best-known work, Dialogue Concerning the Two Chief Systems of the World, *Galileo staged a four-day conversation among three intellectuals. The frontispiece to the first edition pictured those men as Aristotle, Ptolemy, and Copernicus (at right, holding a representation of his Sun-centered cosmos).*

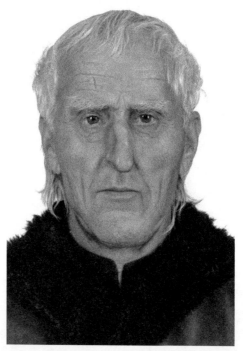

Working from bone fragments, Polish forensic artists imagined the face of their famous countryman as a man of seventy years—Copernicus's age at death.

A Note on the Author

Dava Sobel is the bestselling author of *Longitude, Galileo's Daughter, The Planets,* and *A More Perfect Heaven.* She is coauthor of *The Illustrated Longitude* and translator and editor of *Letters to Father.* She lives in East Hampton, New York.